D0209454

100

YEARS OF PROGRESS

100

YEARS OF PROGRESS

THE OREGON
AGRICULTURAL
EXPERIMENT STATION
OREGON STATE
UNIVERSITY
1888-1988

Oregon Agricultural
Experiment Station
College of Agricultural Sciences
Oregon State University
Corvallis, Oregon
1990

Acknowledgements

Published in 1990 by
Oregon Agricultural Experiment Station
Oregon State University
Corvallis 97331
in cooperation with
Agricultural Research Foundation
Oregon State University
Corvallis, Oregon

Station Miscellaneous 89:3
Printed in the United States of America

Ronald P. Lovell, Andy Duncan, and Richard L. Floyd, Major Contributors
Leonard Calvert, Andy Duncan, Gwil Evans, Richard Floyd, Wilson Foote,
Tom France, Jim Leadon, Readers
Carol Savonen, Editor
Tom Weeks, Designer
OSU Archives, Photographs

Library of Congress Cataloging-in-Publication Data

Oregon State University Agricultural Experiment Station
100 Years of Progress: The Agricultural Experiment Station, Oregon State University, 1888-1988
p. cm.
Bibliography: p. 144
Includes index.
ISBN 0-9622925-0-8
1. Oregon State University. Agricultural Experiment Station--History. 2. Agriculture—Research—Oregon—History. I. Oregon State University. Agricultural Experiment Station. II. Title. III. Title: One Hundred Years of Progress.
S541.5.0720745 1989
630′.720795′34—dc20 89-15994

In grateful appreciation to the Agricultural Research Foundation,
Oregon State University

Contents

Foreword

Agriculture is critically important to Oregon. That's clear to anyone who examines the state's economic foundation. Shifting markets, weather, and numerous other factors make agriculture a quickly changing industry. Research support is important to agriculture anywhere; in Oregon it's vital. We have diverse geography, diverse geology, a diverse climate, and therefore a great variety of crops—and a great need for diverse research.

The Oregon Agricultural Experiment Station, based at Oregon State University, supported the state's agricultural industry throughout the last century. Its story follows. It's an impressive story—of cooperation among farmers, ranchers, other agri-business people, the university, and local, state and federal government—a story of people working together.

History documents that public investment in the Experiment Station's research has paid off for Oregonians, whether they are involved in agriculture or not.

In the early days, several of my predecessors who served as this institution's presidents also served as directors of the Agricultural Experiment Station. The story of Oregon's agriculture and the story of Oregon's agricultural college were frequently synonymous. That's a pretty good indication of the key role the Experiment Station played at this university during the last hundred years. Although today the story may be different, the importance of agriculture during the second century will likely be much the same. Oregon State University is proud of its agricultural heritage; it looks forward to the next century in helping farmers do a better job of producing foods for the peoples of Oregon, the United States, and the world.

John Byrne
President of Oregon State University
1989

Preface

This book was compiled to commemorate the one-hundredth anniversary of the Oregon State University Agricultural Experiment Station.

The sources for this history include: OSU College of Agricultural Sciences departmental histories, early Experiment Station meeting minutes, early Experiment Station quarterly magazine, *Oregon's Agricultural Progress,* Experiment Station Special Reports, history books, and personal interviews.

At the end of most chapters in *100 Years of Progress,* there are a number of excerpts from other publications. Some are narratives; others are chronological listings of events. In journalism, such items are called *sidebars.* These sidebars are included to help make the Agricultural Experiment Station's history come alive.

This project could not have been undertaken without a generous grant from the Agricultural Research Foundation, a non-profit, scientific, and educational corporation which supports Oregon agriculture in close partnership with the OSU Agricultural Experiment Station.

CHAPTER 1

The Hatch Act and Founding of the Station (1888)

Agricultural research programs that began in Oregon and elsewhere 100 years ago provide a rare example of a Federal Government program that not only is so good in concept and execution it can work largely unchanged from its inception, but also provides tremendous good to a maximum number of people at a relatively low cost.

The Oregon Agricultural Experiment Station owes its existence to Federal legislation and funding. The forerunner to Oregon State University, Corvallis College, was designated "the agricultural college of the state of Oregon" by the Legislative Assembly in 1868 so it could accept a grant of land under provisions of the Morrill Act of 1862. In a similar fashion, the Station was set up under provisions of the Hatch Act of 1887.

The Morrill Act in Oregon

Institutions like Corvallis College were obligated under provisions of the Morrill Act to provide instruction that included agriculture, along with mathematics, English, natural science, languages, military exercises, and moral philosophy. These subjects would be added to the classics-based curriculum of Corvallis College, founded by the Methodist Church in 1858. Such changes took time and money, however, and it was more than a decade before they were carried out.

One of the Morrill Act requirements was to establish an experimental farm and buy the appropriate equipment to run it. In 1871, Benton County residents donated a 35-acre farm for the purpose. But state support for the College was slight, and it lived in a perpetual fiscal crisis. Teachers were often unpaid and supported themselves with other jobs.

From 1868 to 1885, Corvallis College was both a private liberal arts college and the agricultural college of the state. The College received on average slightly more than $2,000 annually, in addition to interest from funds obtained by the sale of college land from the Morrill Act. Improvements were few because little money remained after expenses had been paid.

The consequences for research were severe, as noted in an early report:[1]

> This decided lack of interest in scientific agriculture and insufficient appropriations for work, other than the work of instruction in the class-room, has been no doubt due to the newness of the country; its great fertility; exceptionally fine climate, in the farming portion of the state, giving certainty to crops; freedom from insect pests and the extensive tracts of grain and grazing lands to be occupied practically without cost.

In 1885, the church relinquished its claim on funds of the College, and the state assumed control of the institution. The legislature then provided for the permanent location of the state agricultural college at Corvallis if county residents would erect brick farm buildings to accommodate the college at a cost of not less than $20,000. The deadline of four years was met, and the building now known as Benton Hall was dedicated.

After the state took control, the name of the institution was officially changed to Oregon Agricultural College, also called State Agricultural College.

"The general government of the said College shall be vested in and exercised by a Board of Regents to be denominated the Board of Regents of the State Agricultural College of the State of Oregon," read the law passed by the legislature.

When Benjamin L. Arnold succeeded William A. Finley as president in 1872, he sought almost immediately to justify the College's name as a Land Grant institution. He complained about lack of funds and soon got a $5,000 appropriation from the legislature.

He sought to establish both classical and scientific courses beyond those that would narrowly apply to agriculture. Indeed, students majoring in agriculture were part of the physical science school. This in turn, was part of the Scientific Department, which also included schools of mathematics, engineering, practical mechanics and technology, and moral science (political science, ethics, logic, social science). There was also a Literary Department, divided into schools of ancient languages, modern languages, history, and literature.

Agricultural students were expected to complete courses in chemical analysis, chemical physics, natural philosophy, biology, political economy, social science, logic and mental philosophy, English grammar, bookkeeping, mathematics, and military science.

They were also expected to study general methods of preparing soils, the makeup and use of fertilizers, drainage, and the nature and constitution of plants. President Arnold believed that agricultural chemistry was the only scientific foundation for agriculture.

This basic curriculum was later expanded to include courses in entomology, forestry, farm engineering, grasses, landscape gardening, and horticulture.

If the course work for agricultural students was being firmly established, agricultural research was not. Most of the colleges endowed under the Morrill Act, including the Oregon Agricultural College, were doing little or nothing to carry out scientific investigation and experiments in agricultural science.

In an 1884 report, President Arnold cited the work of experiment stations in Europe and noted that a bill proposing the establishment of experiment stations was then before the U.S. Congress. He recommended that the Oregon Legislature send a memorial to Congress urging passage of the bill, which he felt was highly favorable to Oregon interests. The legislature did not act on his recommendation.

Two years later, Arnold tried again.

The Hatch Act is Passed (1887)

Oregon was not unique in the absence of sound agricultural research. Although the improvement of farming through science had long been advocated in the United States, the idea, scorned by some tradition-bound farmers and often misunderstood by the rest, never took hold.

A few scientists, inspired by European examples, nurtured the notion that the systematic discovery and application of scientific knowledge would benefit agriculture. In the last quarter of the nineteenth century, these proponents of public support for agricultural research succeeded in tying their proposal to more general efforts to improve farming and the farmer through education.

Their efforts culminated in national legislation, which in 1887 created a unique Federal-state partnership of aid to agriculture.

In the decade before the Hatch Act, states like Connecticut, California, North Carolina, Massachusetts, New York, New Jersey, Ohio, Tennessee, Alabama, Wisconsin, Maine, Kentucky, and Vermont provided public funds directly to agricultural experiment stations.

Economic depression and disinterest on the part of the U.S. Department of Agriculture stalled any national movement in that direction until the early 1880's. Then, enough states were struggling with the problems of sustained support for experimentation to revive sentiment for a concerted national approach.

A new Federal Commissioner of Agriculture, anxious to gain the backing of the Land Grant colleges for his department, gave experiment station advocates a boost when he called a convention of agricultural college and farmer organization delegates to meet in Washington in 1882. Although agricultural education was the planned focus of the convention, agricultural experimentation was also a concern, especially among the five colleges that had formally established experimental stations on their campuses.

After declaring that scientific investigation was a necessary complement to agricultural teaching, the 1882 convention stopped short of urging that experiment stations that were part of colleges be set up. Instead, delegates endorsed a proposal by the Commissioner of Agriculture that his office serve as coordinator of joint research efforts among stations and colleges. He was entrusted with the task of convincing Congress to subsidize the new approach.

At another series of meetings in Washington in 1883, delegates were asked to endorse a pending congressional bill to create a national system of agricultural experiment stations.

The proposal, authored by Seaman A. Knapp, an Iowa State College agriculture professor, and introduced by Iowa Representative Cyrus C. Carpenter in May 1882, called for providing $15,000 annually from the national treasury to each state agricultural college to operate an experiment station to pursue research into a number of broad areas.

The Federal money would go toward the salaries of scientists and support staff and the expenses of investigation. The colleges would furnish buildings, land, and other facilities.

The colleges would select station directors and staff, who would, in turn, decide on specific experiments to be performed. The general character of this work, however, was to be determined by the station superintendent, the college president, and the Federal Commissioner of Agriculture. By retaining ultimate authority over funds, the commissioner could shape station programs to meet national concerns.

The convention endorsed the proposed legislation and selected a five-man committee to lobby the bill in Congress. The bill was revised to tone down the imposition of Federal authority on colleges and was resubmitted in July 1884. The new bill made stations distinct departments under the colleges and placed their control with college trustees acting through a director and scientific staff. The Federal role was limited to furnishing forms for data tabulation, setting standards for fertilizer valuation, and suggesting lines of inquiry if so requested by a station.

The bill languished in Congress until 1885, when another meeting of agricultural educators in Washington reignited interest. This support attracted the attention of the House Agriculture Committee, which had at least seven other station bills under consideration by 1886. Its chairman, William H. Hatch of Missouri, recommended to the full House in March a bill incorporating the college committee's justification as a preamble. An identical bill sponsored by Senator James Z. George of Mississippi was brought to the floor of the Senate for debate in January of 1887.

The Hatch Act, signed by President Cleveland on March 2, 1887, promoted "scientific investigation and experimentation respecting the principles and applications of agricultural science" through annual grants of $15,000 to each state and territory, to establish agricultural experiment stations under the direction of Land Grant colleges.

Seen as a way to advance agriculture in a rapidly industrializing nation, the Hatch Act created a structure of federated yet independent research institutions to solve farm problems particular to their states while building a core of basic scientific knowledge related to agriculture.

Knapp's original objective that each experiment station was "to conduct original research or verify experiments...[on subjects] bearing directly upon the agricultural industry of the United States" was incorporated into the final act signed by President Cleveland.

The stations were to be "established under the direction" of schools of agriculture in those institutions founded under the provisions of the Morrill Act. In deference to the existing structure of the Connecticut station, however, a state could choose to apply its grant to stations unconnected to its agricultural college. The act also left to the stations the task of determining programs of investigation. The Federal Commissioner of Agriculture's responsibilities were limited to indicating subjects of potential interest, supplying forms, and encouraging uniformity of methods and work.

"By providing funds, a suggestion for organizational structure, and discretionary authority in designing research programs around local needs, the Hatch Act presented an opportunity to benefit agriculture through the application of discoveries of scientific investigation," writes Norwood Allen Kerr in *The Legacy, A Centennial History of the State Agricultural Experiment Stations*. "Over the next century agricultural scientists, land-grant college administrators, and United States Department of Agriculture officials labored within this framework to evolve an agricultural experiment station system which would realize the promise of the 1887 legislation."

Wheat in Oregon

Wheat Came in Covered Wagons

By E. R. Jackman[2]

The Illinois family in 1840, seized with "Oregon fever," took a year or two to prepare for the long overland journey. Clothes were mended and strengthened. Lists were made and remade. Cherished houseplants and unneeded clothing were given away. At last a farm auction was held and a thousand things—"articles too numerous to mention"—were sold to the highest bidder, whose bids were often accompanied by sobs from the farm women who could scarcely bear to see a cherished possession pass into alien hands.

But at last, in the spring of 1842, the family was ready to start. They had one or more strong wagons pulled by strong, patient, sturdy oxen. Oxen, rather than horses, because they did not stampede nor stray so readily; Indians didn't try to steal them; their feet stood the long trail better; and, in the case of extreme hunger, they were better for eating.

Stowed neatly into the wagon, with a stern eye for space, were all the possessions needed to make a new home in the wilderness. A few tools, a few cooking utensils, clothes for different kinds of weather, and the bedding, always including a feather bed, with the load arranged so that the bed could be spread out on top of the load to accommodate a sick owner—if "trail fever" struck him. (The diaries that year and next had daily entries, "passed seven new graves today," "counted only four new graves today.")

But there was one other thing in every wagon, the one thing more essential than all others—food. And it was three-fourths flour and wheat. There was bacon, beans, salt pork, coffee, and sugar. But it was flour and wheat that was hoarded and kept, even if the hardships made it necessary to throw away clothes, bedding, and pots and pans. Flour to prepare in some form every day. Wheat to carry on one's back, if necessary, and to eat, kernel by kernel, as the last resort.

So wheat was foremost in the thoughts of every immigrant. He ate it on the way and kept some, if he could, to go into the ground first when he finally arrived. Their talk was on wheat, and most of the diaries carefully noted the prices charged for flour at the remote trading posts that sprang up along the route. At Fort Hall, flour was $20 a hundred; near La Grande, $40 a hundred.

Wheat Was Once Legal Tender in Oregon

The place of wheat in the economy of the new land was striking. In 1845 a territorial law was passed setting up how debts might be paid, because a man might be considered prosperous, but he might not have any money. The law stated that for payment of taxes and satisfaction of court judgments, the following were legal tender: gold, silver, treasury warrants, approved orders on solvent merchants, and wheat.

The wheat was mainly grown in the Willamette Valley. Right from the first the settlers knew that eastern Oregon would grow it, but there wasn't much local market for it in eastern Oregon and it was too expensive to ship it down the river or over the Barlow trail. Freight charges in 1865 by river and portage, from Portland to Umatilla, were $45 per ton, which prohibited much commercial agriculture.

By 1863 this condition began to focus attention upon railroads and there were many local promoters with grand schemes. A road was building to San Francisco and it reached there in 1869. It was almost intolerable to Oregonians that California, a Johnny-come-lately so far as they were concerned, would have a railroad when Oregon had none. This situation would force Oregon to send to California for all of their eastern supplies, and all mail would have to go that way.

So, in 1863, a public subscription provided money to make a thorough feasibility survey for a railroad to connect Portland with the new railroad at some point in California. The interesting thing is that nearly all of the subscriptions were in wheat.

Oregon Wheat Nourished the Miners

In the meantime, the gold rush had occurred. The mines were pretty much concentrated in northern California and there was also considerable mining in southern Oregon. For years these mining settlements were provisioned by farmers in the Rogue River and Umpqua valleys, and also by farmers near

Portland. Ships came up the Umpqua as far as Scottsburg, bringing supplies from Portland and from the East. From there, long trains of pack mules set out for the Oregon and California mines, but on their return trips they carried loads of flour from the mills in the Rogue River Valley. The ships took this flour to California points where it was again packed in to the mines. The going price in the California mining towns was $25 a barrel for Oregon flour.

So the growing number of flour mills in the interior of Oregon helped to make life tolerable in the mushrooming mining towns of California.

Gold also stimulated wheat growing in eastern Oregon. Held back by lack of market at first, the gold strikes on the John Day, the Burnt River, and on Powder River and Eagle Creek in Oregon, and those in Idaho at the same time, quickly built small cities at such places as Canyon City, Granite, Auburn, and Silver City. So finally there was a wheat-consuming public and in 1870, three acres of wheat were grown near where Weston is now.

By 1890 wheat was a bonanza crop and in 1893 Umatilla County grew 4,500,000 bushels. Of course, the railroad had reached Portland in 1883 and the wheat began to flow to the ports of all nations.

There then followed a period when mills were established to turn the wheat into flour.

Trade in Oregon Wheat and Flour

Before the mills were common, flour was brought to the fur traders by ships from Chile. These ships took on hides in California and furs in Oregon. But with the discovery of gold in California, the Oregon flour mills entered a 20-year period of prosperity. The mines would take all the flour they could send and it was shipped by water from Portland and Scottsburg (Douglas County) and by mule pack train.

Markets soon extended and in 1867 Oregon flour was reported as the highest priced and best flour on the New York market. The writer of this report claimed irritably that "since the boats came by way of California, this flour usually sold as California flour." In 1866 a writer said that in that year $149,065 worth of flour moved from Portland for California, where it sold for $5 a barrel. But at the mines in California markets, it brought up to $25 a barrel.

The earliest recorded export was a report by Avery Sylvester, in command of a ship from Boston. He tells of shipping flour and other products in 1845 from Astoria to Boston by way of the Hawaiian Islands. Trade with Hawaii was rather common. Islanders even made up part of the crews of the fir brigades, notably Peter Skene Ogden, and Oregon's Owyhee River was named after the islands, since Owyhee is nearer to the island pronunciation than our word Hawaii.

In the 1860's wheat furnished the main cargo for many boats operating on the Willamette. In December, 1861, a tragedy occurred. The crop of wheat had been excellent and every warehouse from Eugene to Portland bulged with it. Then the rains started, and the water rose and rose. It carried away some mills and some warehouses, including the large mills at Oregon City. But it swirled into the others, the wet wheat swelled and the warehouses burst. The remaining wheat heated and sprouted and the river boating was a dismal, unprofitable business that year.

Wheat for Export

The new Oregon country desperately needed money. Settlers bought clothes, coffee, furniture, spices, household and farm tools and utensils—and money for these was sent to the East. They mostly started with nothing, and there was an immense need for stoves, for example. Flour, guns, harness—everything took money. So wheat furnished the money.

The "Commercial Review" (Portland) says that in 1868 "the little bark Helen Angier is credited with carrying the first shipment of wheat overseas from Portland." This was evidently successful because within two years, wheat shipments were an established fact. The bark Sallie Brown came from New York around the "horn" in 170 days and returned with a full cargo of wheat and flour. In July, 1870, Robert Meyer and Co. loaded the German ship Herman Dokter with 11,468 centals of wheat. This ship was from China, and illustrates a common route. Atlantic Coast ships carried goods to China and returned via Portland and San Francisco. In that year, 12 vessels cleared from Portland with 242,579 bushels of wheat.

The Willamette Valley was then providing the oil for Oregon wheels of commerce by growing wheat. But by 1870, the valley was too crowded, too conservative, for the restless spirits. Gold had lured some, but the rich strikes were getting scarce, and many miners in what is now Baker and Grant counties, found that long days with a shovel dulled the glitter of gold. These men began to stand up, look around them, and they liked what they saw. They promptly homesteaded.

So settlement of eastern Oregon grew rapidly from 1870 on. Many came, again in covered wagons, but this time the wheels rolled from west to east. There are still a few of these old wagons parked under trees on eastern Oregon ranches. One is on the Locey ranch near Ironside in Malheur County.

But, as noted, freight charges down the rivers were high and the early comers mainly grew cattle and sheep because animals could walk to market. From 1870 to 1883 wheat pressure built up behind the Cascades and some wheat trickled through to the waiting export markets by barge, portage, wagon, river boat, and mule back. In 1883 the Union Pacific reached Portland and the dammed-up wheat burst through.

In the meantime, the mining towns were dying of malnutrition, and the profitable Willamette Valley and southern Oregon trade with the mines slowly petered out so that by 1890 there wasn't much left of this business. Wheat never again dominated the western Oregon business life or farm life. Farmers turned to fruit, sheep, dairy cows, and anything else that would grow in the mild climate.

Chapter 1 Notes

[1] Quoted in Richard Floyd, unpublished history of the Oregon State University College of Agricultural Sciences, 1983 (OSU, Agricultural Communications, Corvallis, Oregon), p. I-2.

[2] E. R. Jackman. IN: Wheat Industry Conference Committee Reports, Feb. 19-20, 1957, Portland, Ore., pp. 1-7.

CHAPTER 2

Early Years of Experimentation (1888-1900)

Oregon greeted the passage of the Hatch Act on March 2, 1887 with elation. The provision of $15,000 a year in Federal funds for the support of the Experiment Station in Oregon increased the total operating budget of the entire Oregon Agricultural College by more than 50 percent. Federal support would increase periodically afterwards.

Oregon Agricultural Experiment Station is Born

Although the Hatch Act had provided funds for agricultural research to the College, it was nearly two years before Oregon passed the necessary enabling legislation. This was partly because some at OAC felt that framers of the original law had failed to consider the need to coordinate experiment station research with resident instruction. On February 25, 1889, Governor Sylvester Pennoyer signed legislation establishing an Agricultural Experiment Station at OAC under provisions of the Hatch Act.

Using the 35-acre college farm as a base, college President Benjamin L. Arnold and his staff began in 1888 to build what would become the Oregon Agricultural Experiment Station. Edgar Grimm, the first Station director, traces origins of the Experiment Station and its ambitious program to start scientific studies in the first Station Bulletin in 1889:

> The farm had to be brought into suitable condition, quarters had to be provided and apparatus purchased before the active work of experimentation could be commenced. This work has been prosecuted with vigor.
>
> A portion of the farm has been tile-drained; soil thermometers planted over the field, as a first step toward a study of the physical properties of the land. The chemical laboratory is being equipped and is nearly ready for active work. While this department is not as well equipped as might be desired yet it is prepared to do valuable work for the station and will grow as the departments and needs of the station may demand.[1]

In 1889, the college farm was increased by 155 acres; farmers and ranchers donated Jersey, Polled Angus, Durham, and Hereford cattle to establish research herds. Advice was going out to farmers to spray trees with paris green and to bend the trees to control the codling moth.

By 1890, Station publications were covering a wide range of subjects, most of them centered on helping farmers solve problems of raising hogs, controlling weeds, irrigating pastureland, improving soils, selecting the right varieties, and constructing useful buildings.

Early Station publications contained both descriptive and experimental information. Some of them were written primarily to answer letters and other queries from farmers. The Experiment Station staff was small but prolific. During the first 10 years, seven staff members

George Coote and his horticulture class at Oregon Agricultural College in 1892.

produced 58 bulletins and circulars. Station Bulletins 2, 4, and 7, prepared by E. R. Lake in 1889, were on horticultural topics of direct concern to growers. The first strawberry bulletin appeared in 1891.

From 1890 through 1907, college presidents were also the directors of the Agricultural Experiment Station. Arnold was first, serving two years until 1892. Others were John M. Bloss (1892-1896), Henry B. Miller (1896-1897), and Thomas M. Gatch, whose tenure in office from 1897 to 1907 bridged the change into the twentieth century.

The fortunes of tiny Corvallis College were indeed on the rise. From its early, nearly penniless days, the small College had been linked to Federal and state governments and been assigned important new tasks that were constantly expanding, according to James W. Groshong in *The Making of the University, 1868-1968.*

Agriculture was the guide and major building block for the institution. Without the Land Grant designation and the funds that went with it, and the later passage of the Hatch Act, Corvallis College probably would have continued to exist in a precarious state and, eventually, died. With agricultural support, the College emerged into a new world of teaching science, technology, and agriculture while continuing the classical education that had, until then, characterized higher education in the United States.

The Agricultural Experiment Station became an integral part of the College, and the work of its staff members began almost immediately to help farmers and other citizens in the state. The College and its fledgling Experiment Station were becoming major forces in the lives of Oregonians.

Minutes of Early Council Meetings[2]

Minutes of early meetings of the Station Council reflect day-to-day operations: organizing a library, getting mail delivered, funding departments, even subscribing to periodicals.

From the October 1, 1890, meeting: "Director stated that the business before the meeting was to adjust the amounts appropriated to the Horticulture and Botanical Departments. Prof. Irish made motion that money be divided in proportion to the amount asked for by each of the departments. Motion prevailed and money was so divided giving $895.00 to Botanical Dept. and $1505.00 to Horticulture Department."

From the November 11, 1890 meeting: "Prof. W[ashburn] made motion that Encyclopedia Brittanica and agril. reports and certain papers be placed where the faculty and students may have access to them. (Motion carried.)"

From the September 28, 1891 meeting: "Pres. Arnold asked that a list of periodicals taken by various departments with date of expiration of subscription be furnished him tomorrow. He also asked about the advisability of mulberry culture and a trial of other new fruits. This was left to the Horticulturist."

From the January 8, 1892, meeting: "Prof. Washburn moved that the freight agents of each company get one of the transfer companies to bring up all freight addressed to College into the departments...and that the job be let alternately between companies. After some discussion on Express delivery the motion was carried."

From resolutions of the council upon the death of Director B. L. Arnold, February 1, 1892: "Whereas: It has pleased the all wise Father to take from us our esteemed Director Prof. B. L. Arnold and

"Whereas: We find that we have by his death suffered a great loss not only of a worthy Councilor but of a dear friend also be it

"Resolved: That we the council of the Oregon Experiment Station . . . testify to our sincere sentiments at our loss and

"Resolved: That we respectfully tender to the affected family and friends of the diseased our heartfelt sympathy and

"Resolved: That these resolutions be (made a part of) our records and a copy be sent to the family and the press.

"Signed: F. L. Washburn
H. T. French
Moses Craig
G. W. Shaw
Geo. Coote"

The council met as agenda items dictated. For the remaining years of the nineteenth century, its members dealt with all aspects of Station business from setting Station Bulletin topics and publication schedules, to debating the merits of buying a typewriter. It regularly heard reports about the various farmers' institutes held around the western part of the state and about the work of the various departments.

Farmers' institutes were a big success. In 1896, the council passed a resolution calling for more of them to be held and asking for funds from the legislature to pay for them. The resolution noted "the great good done to the station by bringing its workers directly in contact with the farmers and fruit growers themselves, that it is the belief of the council that a much larger number of these institutes should be held in the state annually and that an effort should be made to induce the legislature to appropriate a small sum to pay for the services of an Institute Organizer. . . ."

The institutes were given a high priority by the College as well as the Experiment Station. President Miller, also Station director, accompanied A. B. Cordley and Perriot to a series of typical institutes held in southern Oregon in March 1897: two in Ashland, two in Medford, and two in Grants Pass.

"The first evening at each place was occupied with a talk on Insects-Pests of the Apple, Pear, and Peach illustrated by steriopticon views of the various pests," read the council minutes describing the tour. "The second evening at each place was occupied by President Miller with an address on Industrial Education, followed by steriopticon views illustrating the work of the college and station. As much time as possible was given to the inspection of orchards in the vicinity of the various places visited, especial efforts being made to acquaint the growers with the nature of the pests that were injuring their trees..." A local choir performed musical interludes, and a basket dinner was served halfway through the evening.

Council meetings for the remainder of the nineteenth century dealt largely with housekeeping details involving the institutes, the research work of the department, and the Station bulletins detailing the work.

The first sign of direct Federal involvement in Station affairs (since any control was removed from earlier drafts of the Hatch Act) occurred in a directive to agricultural experiment stations from the U.S. Secretary of Agriculture on February 25, 1899. "In view of the increasing number of enterprises in which cooperation between different branches of the Department and the agricultural experiment stations... is being undertaken, it is desirable that the Department should have a general plan for the inception and organization of such work," read the memorandum. "You are therefore directed to report to this office on or before March 15 next the nature and extent of the operations under your charge involving cooperation with the experiment station..., the financial obligations involved..., and the forms of agreement under which these operations are being conducted...."

The First Station-Affiliated Department is Set Up

All of the early research conducted under the auspices of the Station went on in the academic departments of the College. Beginning in 1869, well before the Station was founded, all instruction and research in agriculture was conducted in the Department of Chemistry.

Such research was further defined in 1883 when Edgar Grimm, later named first director of the Agricultural Experiment Station, was appointed professor of agriculture and chemistry. In addition to teaching agricultural chemistry, Grimm also taught practical agriculture, fruit culture, botany, political economy, and organic and inorganic chemistry. This one-man college of agriculture also supervised agricultural experimental work on two acres of ground with the help of two students—all at a salary of $1,000 for nine months.

The original Station employees also taught in OSU's Department of Chemistry. P. H. Irish was the first Station chemist, with W. D. Bigelow, assistant chemist, E. W. Shaw, professor of chemistry and physics and Station chemist, Dumont Lotz, Station chemist, and John Fulton, as acting Station chemist and longtime head of the department.

Fulton's report to College President John Bloss in 1894[3] gives insight into the nature of the work carried on in the department at the time. According to the early report, research projects included the chemistry and physics of alkali soils and methods of improving such soils, chemical evaluation of cattle feeds of the state, and a chemical study of the grasses and clovers of the state.

Departmental professors also studied the composition of various fruits with particular reference to the elements taken from the soil in which they were grown. Samples of fruits and soils were requested from 75 farmers, but Fulton later lamented, "The importance of these determinations was not apparent to many orchardists as only a few complied."[4]

According to Fulton, research was apparently secondary to the pressure for analytical service for all kinds of samples sent from all parts of the state. Samples that were mostly sent in for analysis consisted of ores and waters but also included agricultural products such as milk, honey, peas, bread, syrup, and butter. The laboratory also provided analytical service for the state food commissioner, who sent in everything from butter and cheese to coffee, buckwheat, and spices. The amount of sample work was nearly overwhelming. Noted Fulton in his report: "Considering the amount of miscellaneous work continually received at the laboratory... we have ample employment for nearly all the ensuing year."[5]

James Withycombe, Station Director and Oregon Governor

By George S. Turnbull[6]

James Withycombe, Governor in World War I, crammed three careers of achievement into one lifetime. He was, in turn, outstanding farmer, agricultural experimenter and teacher, Governor of Oregon.

At 17 he was brought to Oregon from his native England by his parents, Devonshire farmers. Coached at first by his father, he early put his mind to farm problems. In a few years he was cultivating 256 acres near Hillsboro on his own. He studied his land, studied his animals and crops, learned about diseases of plants and animals. His farm prospered immensely, and admiring neighbors copied his methods, to their profit. Through a quarter of a century his home place was a model for farmers all over the state.

Oregon State Agricultural College authorities had their eye on this man who was so successfully applying brains to farming, and in 1898 they lured him to their experiment station. Under his direction for 16 years, it became one of the foremost in America.

James Withycombe, director of the Oregon Agricultural Experiment Station from 1908 to 1914 and Governor of Oregon from 1914 until his death in 1919. Photo courtesy of Oregon State University Archives, 413.

Alfalfa, highly important feed crop, he introduced in Oregon. He sold eastern Oregon farmers the idea of rotating nitrogenous crops, thus saving their soil and making them money.

Farmers came to love this modest, friendly, studious agricultural experimenter, who was showing so many the way to success. Finally some Republican leaders thought a man like that might win them back the governorship; the Democrats had won with Oswald West in 1910. Withycombe, never before a candidate for office, won the nomination in a field of eight and was elected in 1914 by the largest majority yet received by any candidate for Governor. Four years later he became the first member of his party reelected to the governorship.

When Professor Withycombe resigned from the experiment station the Oregon *Countryman* evaluated his contribution: "The initial impulse for nearly every great advance in Oregon agriculture may be traced back to the office of this far-sighted, broad-minded master of his craft."

He was a forceful and constructive Governor. As war flared in Europe, he preached preparedness, and Oregon led the national war effort with a volunteer enlistment of 92 percent of its manpower quota.

The Governor combatted the criminal syndicalism menace of the I.W.W., and Oregon's spruce production, vital to the airplane output, was virtually uninterrupted by the "wobblies" while logging operations elsewhere along the West Coast were irregular. He helped bring Oregon out of the war in surprisingly good condition. He took keen interest in the welfare of Oregon's boys in the armed forces.

Highway building in Oregon took probably its longest leap forward in Governor Withycombe's administration with appointment, as authorized by the 1917 Legislative Assembly, of Oregon's first commission named exclusively to handle road matters. Up to then the Governor, the Secretary of State, and the State Treasurer (which officials now make up the State Board of Control) had constituted, ex-officio, the Highway Commission.

Governor Withycombe, good-roads advocate, appointed a commission composed of Simon Benson, Portland timberman; W. L. Thompson, Pendleton; E. J. Adams, Eugene. When Adams retired after a year, the Governor named Robert A. Booth, Eugene lumberman, to succeed him. The resigning commissioner moved on to a position on the Federal Trade Commission.

The 1917 session voted to double the license fees charged motor vehicles and provide for bond issues to match Federal funds, under the Shackelton Act, to the amount of $1,180,310.55. The legislature also called for the issuance of $6,000,000 bonds for the hard surfacing of certain key highways in the state. At that time, not more than about 10 percent of the road (now Highway 99) between Portland and Eugene was hard surfaced.

In his 1915 message the Governor strongly advocated formation of a state constabulary—a project developed and carried out by Governor Julius Meier 16 years later.

Overwork in those feverish days helped bring on a physical breakdown soon after his reelection, and the Governor died March 3, 1919.

"We have a great state," he used to emphasize. Few contributed so much to its greatness.

Experimental Highlights, 1890-1899[7]

1890 Cooking ensilage proves unprofitable. Fruit varietal tests are started with 96 varieties of apples, 21 of plums, 16 of cherries, 42 of peaches, and 40 of strawberries. The Station reports as commodities "suited to our soil and climate," 15 varieties of potatoes, 5 varieties of corn tested for fodder only, 6 varieties of wheat, and 20 of oats.

1891 A Station publication reports: "It has been satisfactorily proved that the more hardy grasses and clovers will give good returns upon the white land when well drained and thoroughly cultivated." Says another: "It was recently well established that it is not profitable to feed whole oats or wheat when fattening pigs."

1892 Experiments with the "codlin" moth continue, and a mixture of IXL, water, and paris green is found effective, "calling for fewer sprayings than anything tried heretofore," says one publication. Woolly aphis [sic] are reported as one of the worst pests. A control method for the grain beetle is found.

1893 Nine farmers' institutes are held. The need for a college dairy plant is reported. Cattle feeding experiments are carried out because, explains a Station publication, "the day of range-fed beef is fast passing away and staff-fed beef is growing in demand." A College spraying outfit is exhibited at the Chicago World's Fair. Acting chemist John Fulton starts a chemical study of grasses and clovers.

1894 Lard retails at 15 to 17½ cents per pound. Wheat is selling at 40 to 60 cents per bushel. A bushel of wheat produces more than 12 pounds of pork in live weight. "These figures would indicate that Oregon can successfully compete with the corn-growing states," reports a Station publication. An IXL spray material consisting of dry lime, sulphur, and salt is recommended to be added to paris green spray used for apple worms. George Coote, horticulturist, recommends careful selection of fruit tree varieties to insure pollination.

1895 "The raising of red clover has ceased to be an experiment with us," says a Station Bulletin.... "It will become much more common [in the Willamette Valley] when its merits are more thoroughly known." Bordeaux mixture is recommended as a spray for apple scab. F. L. Washburn, entomologist, describes a method of making "lime sulphur and salt wash" in two mixtures to be united just before using. Says another Station publication: "The value of a mere chemical analysis of a soil is most doubtful.... It doesn't show how much of a plant food is available."

1896 A. B. Cordley, the new Station entomologist, points out the need for careful laboratory and field work to study the life history of injurious insects. Cordley, Director Miller, and U. P. Nedrick, botanist and horticulturist, join forces to issue a comprehensive bulletin on prune growing, which they call, "the most important orchard industry in Oregon." Flax cultivation is advised by H. T. French to replace grain growing, which he described as "no longer highly remunerative."

1897 Nearly 200 varieties of grasses and forage plants are seeded for test purposes. F. L. Kent reports on rations for dairy cows. Cordley reports a successful spray program for the peach-twig borer.

1898 James Withycombe, vice director of the Station, reports to Director and College President Gatch that "the growing of fiber-producing plants has been dropped for the simple fact that it has been fully demonstrated that a good quality of flax and hemp can be grown in Oregon." Experiments with spraying for the hop louse are conducted. Experiments are carried out using prunes affected with brown rot for the manufacture of brandy, resulting in a product "that would not be called first class" but "probably equal in quality to much that is on the market."

1899 Cordley spends a six-month leave of absence at Cornell University, where he works out the life history of apple tree anthracnose, then locally known only as canker or dead spot.

Chapter 2 Notes

[1] Quoted in Richard Floyd, unpublished history of the Oregon State University College of Agricultural Sciences, Agricultural Communications, OSU, Corvallis, Oregon, 1983, p. I-5.

[2] Early Agricultural Experiment Station Meetings, on file in Director's Office, Agricultural Experiment Station Office, Oregon State University, Corvallis, Oregon.

[3] Quoted in unpublished history of the Oregon State University Department of Agricultural Chemistry, 1965 (on file at OSU, Agricultural Communications, Corvallis, Oregon), p. 1.

[4] Quoted in unpublished history of Agricultural Chemistry, p. 2.

[5] Quoted in unpublished history of Agricultural Chemistry, p. 2.

[6] George S. Turnbull, *Governors of Oregon.* (Portland: Binford and Mort, 1959), reprinted with permission.

[7] Adapted From "The First Fifty Years of the Oregon Agricultural Experiment Station, 1887-1937." Station Circular 125, August, 1937, Oregon State College, Corvallis, Oregon.

CHAPTER 3

The Station Expands (1900-1916)

A Victim of Its Own Success

The new century brought expanded work and responsibilities for Station scientists. In some respects, researchers were becoming victims of their own success. As their laboratory work resulted in improvements in agriculture, they were soon inundated with requests from farmers for information.

The Station chemist, Abraham L. Knisley, for one, tried to deal with such requests himself. In retrospect, it is difficult to see how he got anything else done. Many of his letters, some of them handwritten, were detailed and specific:[1]

> January 18, 1902
>
> Dear Sir:
>
> Your communication about fertilizers for onions has been referred to me. Soils best adapted for onions are usually rich in organic matter and generally respond to applications of potash and phosphoric acid. If it could be easily obtained I would advise using from one-half to one ton of wood ashes per acre broadcast. If this is not obtainable use muriate of potash at rate of 150 to 200 lbs. per acre. . . .

In Oregon, as elsewhere, Experiment Station officials were discovering that they had to develop a way of working with farmers and their families, in their homes. The Station made its first direct link through the farmers' institutes it held in the western part of the state. Staff members like Knisley were responding to individual requests for information. Bulletins detailing research results had been published periodically since the 1890's. But the staff could not pursue research, publish results, and maintain direct personal relationships with farmers, all at the same time.

Farmers were busy people, too. They seldom took vacations and, except for the small number who lived nearby, could not spare the time to travel to Corvallis to get the information they needed.

Not all the problems of the farmer could be solved by mail, however. A way had to be found to take the information gained in laboratory research directly to those who needed it. The farmers' institutes held in Oregon and around the United States were one answer. By 1902, 2,700 institutes around the nation had attracted 800,000 farmers and their wives. These institutes eventually led to the establishment of the Cooperative Extension Service. The Oregon Extension Service was established in 1911, and the first county extension agents began work the next year in Marion and Wallowa counties. In 1914, the Smith-Lever Act gave major congressional support to extension work throughout the United States.

In 1900, the Experiment Station began cooperative work with the U.S. Department of Agriculture on the experimental growing of sand-binding grasses at Gearhart Park on the north Oregon coast. That same year, the summer fallow system of wheat growing in the Columbia Basin was denounced by Knisley, the Station chemist, as "suicidal." The first formal course in veterinary medicine was taught to nine students over three terms.

Experiment Station scientists were busy. In a 1901 report they were described as follows:[2]

> The Station staff for the past year has worked mainly on practical subjects which were of immediate pecuniary interest to the agricultural classes. Examples of these are indicated in the chemist's report of his work on the evaporation from fruits, chemical study of silage and oil analysis; and in the report of the entomologist of his work with curly leaf, bacterial disease of strawberry plants, and fungus disease of wheat.

Union, the First Branch Station

Because of its federal relationships, the Experiment Station was becoming aware of its broader responsibilities. The members of the council realized that the Station could no longer focus on just the Willamette Valley. Across the Cascade Mountains was another Oregon, a high, arid land unsuited to the kind of agriculture that flourished in the Valley.

In 1901, the Oregon Legislature appropriated $10,000 to establish the first branch station at Union in northeast Oregon. Consisting of

600 acres of fertile but poorly drained soil, it was located on the west edge of Union. The original brick office, built in 1901, is still used today, although it has been modernized with electricity, indoor rest rooms, and a central heating system.

One of the early objectives of the station was to help ranchers drain, irrigate, and improve valley land for forage and livestock production. Much of the original tile drainage installed before 1910 is still in use. Additional objectives were to provide superior cattle, sheep, swine, and draft horse breeding stock, to test varieties of grasses, legumes, and cereals, and to provide seed for improved varieties. Considerable time was also devoted to vegetable and berry growing for the home gardener.

Eventually, a whole network of stations would be established across the state to deal with its widely dissimilar agricultural regions. These branch stations and the research carried out on the main OAC campus in Corvallis would come to have an almost incalculable impact on Oregon agriculture. Experimental work under the Hatch Act would produce new varieties of vegetables and fruits and improved meat animals and fowl. This research also resulted in new and more economical farming methods and better machinery and equipment. Most pests and weeds were either eliminated or controlled.

Kerr Strengthens OAC and the Station

The Experiment Station also had a direct effect on higher education. By providing support independent of teaching budgets, Hatch Act money freed scientists for full or part-time research. Because of the Station, research became a regular part of the work of the College.

The acceptance of research as part of the College function was not as widespread as might be expected, however. Strong ideas about what the College should be were firmly imbedded in many minds. "Our college should forever remain as it is—emphatically, the farmers' school," OAC President Thomas M. Gatch had written in a report to the regents.[3]

During the first decade of the 20th century, there was no question that OAC was a "farmers' school" in at least two respects: first, it was so named by Oregon law, and secondly, three-fourths of the population it served still lived on farms. But the institutional doctrine of "education plus training" applied to agricultural students as well as to others.

Students in agriculture were required to study the English language, English literature, speech, history, drawing, political science, and psychology, with additional electives available in language, literature, and science. Gatch himself taught psychology and economics, and at least a third of the 26 faculty members taught subjects having no direct connection to agriculture, engineering, or related fields. The College had become much more than a "farmers' school," no matter what Gatch thought.

Gatch's retirement in 1907 brought to OAC a new president with a much broader view. Although only 34, William Jasper Kerr had headed two major institutions in Utah before accepting the job in Corvallis. When he left for his new assignment in Oregon, the Logan newspaper praised him as Utah's most distinguished educator. He rapidly acquired a similar reputation in Oregon, according to James W. Groshong in *The Making of a University 1868-1968.*

"It is difficult to assess briefly the influence of William Jasper Kerr on the growth and development of the institution," writes Groshong. "Every president to some extent stands on the shoulders of his predecessors. . . .

"Moreover, Kerr arrived at a time when conditions in the state were favorable for rapid expansion, as they had not been under Arnold

Class demonstration of how to dehorn a cow at Oregon Agricultural College, 1909. Photo courtesy of OSU Archives, 939.

or even under Gatch. But it is none the less true that the College and the state had never before had the benefit of such a combination of skill and energy as was exemplified in the new president."

In his first year, Kerr established four major schools at OAC: Agriculture, Engineering, Home Economics, and Commerce. Subsequently, except for the College of Liberal Arts (earlier called Lower Division, then Humanities and Social Sciences before its change to the present name), Veterinary Medicine (which began as a department of the College of Agricultural Sciences), and Oceanography (which began as a part of the College of Science), Kerr supervised the creation of all the major schools which now make up Oregon State University: Forestry (1913), Mines (1913, incorporated with Engineering in 1932), Pharmacy (1917), Education (1918), Health and Physical Education (1913).

In his quarter-century as president, Kerr also supervised the construction of a number of major buildings, among them the Mens' Gymnasium, the Womens' Building, Milam Hall, the library (now Kidder Hall), Graf Engineering Laboratory, Dryden Poultry-Veterinary Hall, Covell Hall, Production Technology, the Armory, and the Memorial Union. In 1907, he came to a 225-acre campus worth $229,000.

William Jasper Kerr, president of Oregon Agricultural College from 1907 to 1932 and director of the Oregon Agricultural Experiment Station in 1907, had a large influence on the expansion of Oregon Agricultural College. Photo courtesy of OSU Archives, P1:9.

In 1932, he left a 555-acre campus valued at $7.5 million. His first annual budget had totaled $88,000; his last reached $2 million.

Kerr conducted his job by a credo he later expressed in a 1931 speech before the presidents of Land Grant institutions[4]: "That Land Grant institution . . . that most fully surrenders itself to the state and nation in a spirit of service, that institution shall truly be greatest among us."

In his 1906-1908 report, Kerr made it clear that in a Land Grant institution the liberal arts should have a limited, largely utilitarian function. He cited Morrill's provision for "liberal and practical education" but saw these as working together "to apply science in the industries of life."

Kerr and other early educators realized that once experiment stations were functioning, it was inevitable that Oregon Agricultural College would have to find a way to work with the farmer and his family. The Station had already established the first direct link with its farmers' institutes.

In the first decade of the 20th century, Oregon had about 45,500 farms. In 1909, railroads had penetrated the vast reaches of central Oregon, and people there looked forward to agricultural development. Not only were OAC faculty members contacted directly about solving problems, but they also were asked to speak at teachers' conferences, local chamber of commerce events, high school commencements, grange picnics, grower association conventions, and other meetings.

Railroads also asked the agricultural faculty to work as part of agricultural demonstration trains. Railroad cars were outfitted with agricultural equipment and even livestock, poultry, and other demonstration material. College personnel would present instructive programs along the rail line. Similar demonstrations were given at the Oregon State Fair, farm organization meetings, and at elementary schools.

Kerr decided that this growing program of off-campus education should be directed by one person. The regents agreed and, as a result the Extension Service was established in 1911 as an administrative division in the School of Agriculture.

OAC Departments Expand

Throughout the Kerr years, individual academic departments were being set up that were largely funded by, and did the research of, the Experiment Station.

Agricultural Chemistry

As noted in Chapter 2, the Agricultural Chemistry Department was founded in 1883. By 1906, under the direction of Charles E. Bradley, Station chemist and instructor in the department, chemistry activities began to include research. Analytical service work was relegated to a position of secondary importance in Station programs.

About 1907, a state fertilizer and agricultural lime law went into effect, and analysis of all commercial samples was made a department responsibility. Horticulture was then assuming a prominent place in state agriculture, and as a result, research on fungicides and insecticides became an important part of departmental research. Scientists carried out an extensive study of the chemical reactions involved in the preparation of lime-sulfur spray, a widely used fungicide and insecticide still commonly used today. These scientists also completed the first chemical investigation of hops made at the Oregon station. H. V. Tartar, research chemist, and F. C. Reimar of the Southern Oregon Branch Station, Medford, researched the importance of sulfur fertilization for legumes. Later, they completed extensive soil analyses that showed a sulfur deficiency in many Oregon soils.

Poultry Science

The Department of Poultry Science was first organized in 1907 with James Dryden as head. The first courses in poultry husbandry were offered the following year. Dryden came to OAC from Utah Agricultural College, where he had served as photographer, meteorologist, poultry manager, and assistant in animal industries for the Experiment Station there. At Utah he also served as private secretary to William Jasper Kerr.

In 1907-1908, 10 acres were allotted to the department, and several buildings were erected, including an incubator house with 16 incubators and 28 small, movable colony houses for laying hens and chicks. Also in 1908, Dryden started chicken-breeding experiments with 59 white Leghorn and 110 Barred Rock pullets. Early incubation experiments were concerned with the effects of humidity and ventilation on hatchability. Experiments on cooling and turning eggs during incubation were also conducted. As early as 1911, Dryden observed that shells of eggs incubated under hens contained more oil than those in incubators.

Dryden's first scientific paper, "Some recent experiments in incubation," was published in *Proceedings of the International Association of Instructors and Investigators in Poultry Husbandry*, 1909. Another department milestone occurred in 1913 when Lady McDuff, hen #C521, completed 365 days of egg-laying with a record 303 eggs. The publicity received by this record focused both national and international attention on the poultry breeding research at OAC. Some in the College felt that the hen had really put the Oregon Agricultural Experiment Station on the map for the first time.

Entomology

Entomology had been taught at OAC as an undergraduate subject since 1873. The Department of Zoology and Entomology was organized in 1889. In 1914, entomology became a separate department.

Agricultural Experiment Station research has been a predominant function of the department since 1888, when F. L. Washburn became the first economic entomologist. He published a number of bulletins, mainly on codling moth and San Jose scale. In 1908, he became dean of agriculture and, in 1911, department head.

In 1911, the department received funding for insect control research authorized by the Crop Pest and Horticulture Law. As a result, scientists researched insect control in orchard, vegetable, and small fruits. At this time they also began the first studies of field crop insects, especially those in vetch, clover, and alfalfa.

Farm Crops

Farm crops was first organized and recognized as a separate department in 1916. The history of instruction and research in farm crops goes back through the Agronomy Department (1907-1916) to the Agricultural Division formed at the time of the establishment of OAC in 1868. During the early years of the College, gardens were maintained for the production of crops. Different varieties and cultural practices were tested and students were instructed in crop production.

The research work of this department and its predecessors reads like a history of Oregon agriculture. All of this work was paid for by the Station and carried out under its auspices.

In 1900, earlier work with vetches and peas was expanded, and trials were started at Moro. Sweet sorghums for syrups were tested. Two cooperative trials with the USDA were started in 1901: testing of sand-binding grasses at Gearhart Park, and testing of red clover samples from different states and countries. Clovers, vetches, and peas were also recommended in the same year for western Oregon.

By 1901, experimental plots of alfalfa had been in production for six years. In 1902, a study of surviving types was made on a plot that had been seeded with a mixture of 33 grasses in 1898. Only nine of the original 33 could be found. Trials with European hop varieties continued.

Hop drying and chemical analyses trials were started by the Department of Chemistry in 1903. Malting barley samples were analyzed for protein content in 1904. In 1905, experiments on liming and seed inoculation with alfalfa continued. Selection of vetch varieties for high protein content was continued. By 1906, experiments had demonstrated that alfalfa could be grown successfully in western Oregon.

The establishment of the Agronomy Department in 1907 increased crop research at OAC; crop rotation, corn breeding, and breeding and testing of winter and spring varieties of wheat, oats, and barley kept researchers busy. There were irrigation experiments on clover, alfalfa, potatoes, corn, beets, and kale, and research to improve grasses legumes, and potatoes. The foundation was laid for the production of improved seed crops, potatoes, wheat, and forage and for seed certification.

Agricultural Engineering

As with many of the academic disciplines that comprise agriculture at OSC, it is difficult to determine precisely when agricultural engineering became a separate department. The 1885 and 1886 OAC catalogs refer to courses in farm implements. The 1893-94 catalog included the word "engineering" under the Agricultural Department, and it reappeared regularly from then on.

Agricultural engineering as a separate subject probably began on the OAC campus in 1910, when a few lectures and laboratory periods were devoted to horse-drawn field machines and steam traction engines.

Research conducted by department staff has been varied. One of the most significant developments in OAC agricultural engineering was in sprinkler irrigation, an approach that eventually spread throughout the United States and abroad. Early investigators devoted a great deal of time to researching the drying of walnuts, filberts, corn, hops, prunes, hay, and ladino clover.

Research involving application of agricultural chemicals by spraying and soil fumigation has been very important because without these techniques, many of Oregon's most fertile fields would be rendered less productive by various soil-borne pests.

In cooperation with the Squaw Butte Branch Station near Burns, an agricultural engineering scientist, Dean E. Booster, developed a seeder to seed sagebrush-bunch grass rangeland. Another scientist, R. N. Lunde, developed a filbert and walnut harvester using the vacuum pickup principle. Department scientists have also conducted research on migrant farm and ranch housing.

As early as 1938, the U.S. Department of Agriculture had a unit working in cooperation with the department, investigating fiber-flax and small-seed harvesting and processing.

Soil Science

Although soils research and teaching at OSC go back as early as 1873, a soils department with its present subject matter responsibility dates back to only 1918.

"Modes of drainage" was listed as a course of study in the 1873-74 catalog. By the 1890's, course offerings included "Origin and Formation of Soils," "Mulching and Management" and "Soil Exhaustion and Methods of Improving Soils." In these early years, the Department offered the first course on irrigation farming in the United States. Soils research before the turn of the century included investigations in subsoiling, irrigation, drainage, and fertilization.

Sulfur was established as a necessary amendment for alfalfa in the early 1900's, and subsequently for many crops in both eastern and western Oregon. Research on crop production on the Red Hill soils of the Willamette Valley established the need for phosphorus and liming.

Departmental cooperation with soil surveying began in the early 1900's; soil surveys of Coos, Curry, and Klamath counties were completed before 1910. Subsequent soil surveys concentrated on proposed reclamation projects in Oregon, with scientists cooperating with the U.S. Bureau of Reclamation to advise settlers on farming practices in specific areas. The Department of Soils carried out soil surveys for the river basin studies of the 1960's and has continued to cooperate with the Soil Conservation Service in modern soil surveys.

Early experiments on reclamation of alkali soils were carried out near Vale and later in Klamath County. Tile drainage of Willamette Valley soils was tested experimentally in 1907. Supplemental irrigation for conditions in western Oregon, where water stress becomes common in late summer, was begun as early as 1907, leading to large increases in production.

With the research of Chester Youngberg, the department has long been a leader in forest soils work, establishing the importance of the land base for the forest industry in Oregon. The unusual characteristics of Mazama ash in influencing the growth of trees led to a detailed study. Use of fertilizers for crop production expanded greatly during the 1950's and 1960's. Thomas Jackson began his career in the department in 1952, studying many of the plant nutrient requirements of soils and crops in Oregon. Intensive use of poorly drained soils in the Willamette Valley was a major departmental project in the 1960's for faculty including Larry Boersma.

Home Economics

About 1910, research was also beginning in the School of Home Economics. "It seemed quite natural to me . . . to use a scientific, investigative approach in teaching the various phases of homemaking," wrote Ava Milam, longtime dean of the school in her autobiography, *Adventures of a Home Economist,* she co-authored with J. K. Mumford. "The accumulated knowledge of the past was not enough; we needed a body of new knowledge. As we took our first tentative steps into graduate work, we thought of it in terms of original investigations to help develop a store of new knowledge."

The Home Economics biennial report for 1914-16, which noted that graduate enrollment had increased 100 percent in the biennium (from 4 to 8 students), stated that "graduate work is offered in household administration, house decoration, textiles, and in research work in foods. . . . Experiments have been carried on to determine practical recipes for the use of English walnuts . . . the relative cooking qualities of different kinds of potatoes . . . and profitable disposal of by-products of the loganberry industry . . . This knowledge, when collected and disseminated, will aid . . . industries of the state to an extent which cannot be estimated."[5]

A beginning in textiles research at that time included a project requested by the Laundrymen's Association of Portland to determine the amount of adulterations in staple dry goods that find their way into a laundry. The school asked for staff time to increase the amount of research work, but many years passed before funds became available for this purpose.

In 1915 the first student in the school completed her master's thesis based on original research. Harriet B. Gardner, a recent graduate of Michigan Agricultural College, came to work for an advanced degree in home economics at OAC and to live in the home of her brother, Professor Victor R. Gardner of the Horticulture Department. At this time, the apple-growing industry was achieving importance in the state's economy, but there was little information available on the cooking qualities of the different varieties of apples produced. So Harriet tackled the problem of testing different varieties of apples in five different food products.

Her brother and another horticulturist, E. J. Kraus, agreed to assist and provided 71 different varieties of apples for her to test. Her study had three objectives: (1) to determine the relative value of different varieties of apples for sauce, pies, dumplings, jelly, and marmalade; (2) to determine some of the general principles underlying these cooking properties; and (3) to ascertain if differences in cooking qualities were associated with differences in gross morphology and cell structure of the fruit.

Using the same recipes each time, Harriet Gardner tested each variety as it reached its prime for cooking from August to April. Each time she finished a batch, five judges scored the products for flavor, texture, color, tenderness, clearness, and so forth. Using averages of scores, she rated each apple variety as excellent, very good, good, fair, or poor for each of the products. Six varieties were found to give good products with any of the cooking methods employed—Maiden Blush, Tompkins King, Jonathan, Grimes, Rambo, and Northern Spy. Other varieties proved excellent for one or more of the cooking methods.

Results of this study served not only the housewife who chose apples to cook, but also the nursery manager and orchardist who selected varieties to plant and the grocer who chose varieties to sell. To give the findings wide dissemination, the Agricultural Experiment Station published the results of this study in a station bulletin in 1915.

Umatilla, the Second Branch Station

In 1909, the U.S. Reclamation Service began the development of the Umatilla Irrigation Project on 25,000 acres in northwest Umatilla County, near the confluence of the Umatilla and Columbia Rivers. In 1917, another 11,000 acres were added. From an engineering standpoint, the plan for impounding and delivering the water was efficient and sound. Unfortunately, the soils to be irrigated were not evaluated before construction began. It soon became apparent that such an investigation would have determined that some of the area was too sandy for furrow or flood irrigation.

Wide publicity about the new irrigation project brought a steady number of new settlers into the area. Many had no previous farming experience and little money left over after they had paid excessive prices for their property. The sandy land they tried to cultivate was often incapable of growing anything.

It immediately became necessary to provide settlers with information to guide them in their efforts to establish farms. A way had to be found to stabilize and increase productivity in the windblown soils. There was also a great need to learn how to handle irrigation water.

The Oregon Agricultural Experiment Station came to the rescue by entering into a cooperative agreement with the Bureau of Plant Industry of the USDA. Together they established local research programs that would furnish the greatest possible assistance to the settlers in reclaiming and farming the low-fertility sandy soils. The Bureau agreed to contribute $3,000 annually for the research if the State of Oregon would give a similar amount. Funds were appropriated in the 1909 Oregon Legislature.

The original Umatilla Branch Station was located on a 40-acre tract of Federal land that had been withdrawn from entry in 1908 by the Department of Interior for use as an experiment station. Through the years, the original site was found to be too small, and much of it unsuitable for the experiments being conducted. In 1931, an order signed by President Herbert Hoover gave the station 180 acres three miles south of the original site, where an enlarged research program was initiated. The original site reverted to the U.S. Reclamation Service.

From its inception, the Umatilla Station tested and evaluated hundreds of crops on thousands of plots under a wide range of irrigation, fertilization, and cultivation practices in order to find varieties best adapted for market and for home use under prevailing climatic conditions.

The Station tested nearly every conceivable method of putting humus into the sandy soil of the region to increase production and reduce wind erosion and irrigation needs.

Good Work Being Done

Brief Sketch of the Progress Made at the Union Experiment Station the Past Year

from *The Weekly Eastern Oregon Republican*
Union, Oregon, Saturday, September 6, 1902

One of the most interesting institutions at Union is the State Experiment Station, says a correspondent of the *Portland Journal.* It is located on the tract of 620 acres of land that was purchased by the state for a branch insane asylum. Failing to secure the branch insane asylum it was turned over to the state a little over a year ago to be used as a branch experiment station, to be conducted in connection with the Agricultural College of Corvallis, Oregon. A neat brick building costing $3,500 was erected on the grounds. When the land was turned over a considerable portion of it was unfit for use for farming purposes, as it overflowed.

Last winter a mile and three-quarters of ditch was dug, with an average depth of three feet and nine inches. It was dug at a very slight cost to the state and has rendered every acre of the land susceptible of cultivation. A considerable portion of the land consists of peat-land, which is considered one of the best soils, as it is almost inexhaustible in its fertility. A soil containing much humus does not wear out readily and contains a great amount of the elements of plant food. At present 160 acres are in use for experimental purposes. One of the most important experiments, locally, is the testing of sugar beets. It is estimated that sugar beets cost from $35 to $40 per acre to raise. This sum includes the rent of the land or the interest on the capital invested, the cost of the seed, the labor of cultivation, and the cost of transportation to the factory. From 7 to 18 tons of sugar beets are raised to the acre. "The farmers do not begin to receive the value of their beets," said A. B. Leckenby, who is in charge of the experiment station. "They receive but $4 per ton for their sugar beets at the factory. The beets contain, on an average, 400 pounds of crystalizable sugar per ton and this does not take into account the secondary sugar or the molasses. Sugar beets could be made a very profitable crop if the farmers erected a co-operative sugar factory or paid the present factory a fair price for converting their beets into sugar instead of selling them outright. . . . Then the present method of cultivation is very wasteful. In the place of drilling in the seeds and afterwards weeding out the surplus plants, they should be sowed in hills. We find that this effects a saving of seven-tenths of the seed. Another point in favor of hill sowing is this: The potash salts or alkali works to the top and forms a crust which a single beet finds difficulty in penetrating. When they are sown in hills a cluster of young beets lifts the crust with ease. Then the system of cultivation with a hoe is drudgery and results in useless loss of time. We use a tool I recently invented, called the scorcher."

To my question, "What is a scorcher, and if it is so good why don't you patent it?" Mr. Leckenby responded:

"I never patent anything and I have invented a good many labor-saving devices. I believe in working for the good of all and not, by patenting an article, placing it out of the reach of those who need it most. The man who lightens the burden of labor is a public benefactor, while he who restricts the use of labor-saving devices is not. Here is the scorcher. It will cost a farmer not to exceed half a dollar and will do better work and three times as much as a hoe with a tithe of the muscular exertion."

Many very simple and effective and easily-made tools are used on the experiment farm. . . .

The experiment farm will harvest large amounts of seed this year of various grains and grasses. Much of it is for free distribution to farmers. There will be at least a ton of Oregon brome grass seed for free distribution. It is a very hardy grass of great nutritive value. Last winter the Oregon brome grass was choked out by mustard and wild oats. The frosts killed the weeds and did not injure the brome grass in the least. . . .

The experiment station is a very valuable thing to Eastern Oregon, for under the competent management of Mr. Leckenby valuable experiments in grasses suitable to arid and semi-arid lands are being made.

Hon. J. M. Church, one of the regents of the State Agricultural College, while in Union visited the experiment station and reports that he is well pleased with the work that is being done under Superintendent A. B. Leckenby. Mr. Leckenby has discovered a native grass that no less an authority than Dr. Withycombe states will prove to be of more value to this state than all the money combined so far expended on this school.

Experimental Highlights, 1900-1916[6]

1900 Cooperative work with the United States Department of Agriculture is started on the experimental growing of sandbinding grasses at Gearhart Park. Field tests made of 16 samples of red clover from different states and countries.

1901 "The Station staff for the past year has worked mainly on practical subjects which were of immediate pecuniary interest to the agricultural classes. Examples of these are indicated in the chemist's report of his work on the evaporation of fruits, chemical study of silage, and soil analysis, and in the report of the entomologist of his work with curly leaf, bacterial diseases of strawberry plants and fungus diseases of wheat." Dr. James Withycombe reports that one of the most serious problems confronting the agricultural industry of the state is the subject of growing leguminous or humus-forming crops in the semi-arid portion of the state, "Land that is producing at present satisfactory crops of cereals will be transformed into desert wastes, if some form of cropping cannot be devised to supersede the bare summer fallow." Growing of rape seed in western Oregon recommended. The first Branch Station is established at Union.

1902 "Professor Pernot, station bacteriologist, has by some investigation been able to can cheese and control its flavor as desired. This is a very valuable discovery. In this production there is no rind, no mold, no evaporation." "We have an abundance of material in this state for fruit and vegetable canning, and promotion of this industry should be taken up by the college and station."

1903 "It has been conclusively demonstrated by this station that alfalfa can be successfully grown in this section [western Oregon]." Apple scab, San Jose scale, and crown gall are the chief problems being studied by A. B. Cordley. Method of perpetuating pure cultures of organisms for butter starter developed by E. F. Pernot.

1904 "Horticultural work is practically nil.... The station suffers a serious embarrassment for the want of a scientific and practical horticulturist." "There is evidently a large and profitable field for irrigation awaiting development in western Oregon." "For several seasons the Station has secured gratifying results from two general systems of crop rotation." Inoculated soil distribution made to farmers and work was started in the development of bacterial cultures for legumes.

1905 C. I. Lewis added to the staff as horticulturist. Director Withycombe advocates broadening the scope of the Station by the addition of work with poultry and apiculture. Feeding of skim milk to hogs found beneficial.

1906-07 Report made by Director James Withycombe to President W. J. Kerr: "Without doubt fruit growing is the most rapidly developing branch of agriculture at the present time. . . . There should be a special representative of the station at Hood River." Cordley, dean of agriculture, starts study of stock solution of lime-sulphur. "The growing of clover, vetch, alfalfa, rape, and kale throughout the western portion of the state, we modestly claim, is almost entirely due to the efforts of this Station. This change from the system of grain growing and the bare fallow has been worth millions of dollars to our farmers." One hundred acres of additional land needed. Duty of water studies started at Hermiston. First demonstration train operated through the Willamette Valley and the Columbia River Basin. E. F. Pernot makes report on tuberculosis as a disease of chickens.

1908-09 "A flock of 200 fowls was purchased from different breeders of the state last fall. Trap nests were installed in all of the houses and a record is being kept of the number of eggs laid by each hen. At the end of the laying year selection from this flock will be made on the basis of eggs laid, and these will be used for breeding next year. This is probably the most important line of work that the poultry department can carry on." James Dryden is in charge of this poultry breeding work. Hermiston Branch Station established January 6, 1909. Union Branch Station studies finishing range cattle for market. Study of apple tree anthracnose advanced.

1909-10 Pumping plant is installed and supplemental irrigation started on alfalfa, potatoes, clover, and corn. Organisms causing cherry gummosis are isolated for first time. Dry-farming investigation is started on the Moro Branch Station. Need of extension department reported to handle "all institute work and at the same time attend to much of the general correspondence of the station."

1910-11 Branch stations are established in Harney Valley and southern Oregon. A. B. Cordley continues detailed study of lime-sulphur compound. More than 100 varieties of vetch tested, becoming a valuable seed crop. Irrigation proved valuable on a number of western Oregon crops. Legume inoculation makes it "entirely practicable to grow alfalfa in the western part of the state."

1911-12 World's first 300-egg hen produced. C521, a hybrid "Oregon" breed hen, is first in the world to complete a trap nest record of 300 eggs in a year. Chemical spray investigations reveal the nature and reaction of lime-sulphur spray. The Dairy Department reports on successful use of kale as a major dairy feed. The Poultry Department reports that "all records of the past three years indicate that by careful individual selection the laying qualities may be improved in any of the breeds."

1912-14 James Withycombe retires as director of the Experiment Station. . . . "Perhaps the most notable value of the experiment station work has been its influence for better agricultural and horticultural practices throughout the state, including its influence of

	better methods of handling livestock, especially the growing and fattening of swine." Strawberry breeding work started. Prune standardization investigations started to "effect a definite standard or standards for the commercially dried prune in the state... including the technique of evaporation, sanitation, chemistry of the product, molds, insect infestation, processing, grading, and packing for market." Potato diseases receive special attention.
1914-16	A. B. Cordley, now dean of agriculture and director of the Experiment Station, reports "The last biennium has been a period of retrenchment in expenditure and of curtailment of work. The Legislature at its 1915 session, by enactment of the law 'to repeal all continuous annual and standing appropriations,' deprived the Oregon Experiment Station of all state support other than a few small appropriations for branch stations. The loss made it necessary to dispense with the services of nine members of the staff, to abandon entirely several lines of investigation, and reduce other work." Experiments in southern Oregon and near Salem revealed that sulfur is a fertilizer of great value for alfalfa. Studies of abortion and sterility in cattle started. Need of a horticultural by-products building urged.

Chapter 3 Notes

[1] Quoted in James W. Groshong, *The Making of a University, 1868-1968,* Oregon State University, Corvallis, Oregon, 1968, p. 13.

[2] Quoted in Richard Floyd, unpublished history of the Oregon State University College of Agricultural Sciences, Agricultural Communications, OSU. Corvallis, Oregon, 1983, p. II-1.

[3] Quoted in Floyd, unpublished history of OSU College of Agricultural Sciences, p. II-2.

[4] Quoted in Groshong, *The Making of a University,* p. 24.

[5] Unpublished history of the School of Home Economics, Oregon State University, Corvallis, Oregon.

[6] Adapted From "The First Fifty Years of the Oregon Agricultural Experiment Station, 1887-1937," Station Circular 125, August 1937, Oregon State College, Corvallis, Oregon.

CHAPTER 4

World War I and the 1920's (1917-1929)

War Clouds Over Europe

Although Europe was far away, Oregon Agricultural College was not unaffected when war broke out there in 1914. Because of conditions brought about by the war, the demand for rennet extract for cheese-making became greater than could be supplied by manufacturers. The shortage was especially severe in the Pacific Northwest. Station scientists conducted a series of experiments with pepsin, which alleviated the shortage and strengthened the reputation of the Station as a problem-solver for Oregon.

On campus, military science had been taught for years, as it was at all Land Grant colleges. As soon as the United States entered the war in 1917, many members of the OAC community—faculty, staff, and students—enlisted and were soon fighting on the mud-soaked battlefields of France and Belgium.

"With records quite incomplete, it is known that 19 members of the faculty and staff of agriculture and nearly 300 students registered in 1916-18 are in military service," Dean Cordley reported.[1] Enrollment at OAC was then about 4,000.

In 1917, the new Horticulture Products Building was taken over as a storage facility for the student Army Training Corps stationed on campus.

War realigned Station goals. Projects on garbage, peanut hulls, coconut meal, molasses, and oat hulls involved practical short term studies. With the national emphasis on food production, Station botanists studied only diseases affecting the great food crops, particularly grain and potatoes.

Staff members joined the Greater Production Campaign and the Seeds Committee of the State Council of Defense and they were advisers to the Public Service Commission and the Bureau of Markets.

President William Jasper Kerr responded when the Secretary of Agriculture telegraphed an inquiry asking for a series of patriotic meetings. The "existing situation as regards agricultural production and war needs" was presented in 30 visits around the state.[2] The meetings apparently had their effect because agricultural production went up and Liberty Loan drive subscriptions increased.

Oregon also responded well to a request from the Secretary of Agriculture that a county agent be established in every agricultural county so the nation's need for food production could be met quickly.

With the good response to national requests and the resultant rise in patriotism, however, came hard times—for scientists, farmers, and students. Reported Obil Shattuck, superintendent of the Harney County Experiment Station in Burns, about the 1918 season: "The results of the best season's work at this Branch Station can be well explained in five words, viz, drouth, frost, grasshoppers, sage rats, and jackrabbits."[3]

The signing of the armistice in 1918 brought new problems to Oregon farmers. Because of the war effort, farmers had geared their operations to high production. A short economic depression in 1920 accentuated their problems. As late as 1919, egg production in Oregon was inadequate for state needs, and eggs were shipped in at times from the Midwest. Dairy production did not meet the needs of the state. In some seasons, butter and cheese were shipped into Oregon by the carload.

By 1923, however, production of both butter and cheese had more than caught up with consumption. Surpluses were accumulating, depressing markets. Producers were at a loss about the direction of their industry. Production was here, they said, but what about distribution? Both the Experiment Station and the Extension Service stepped into the breach, finding new crops and markets, and vowing to learn more about marketing.

The Departments Continue Their Service

Agricultural Chemistry

R. H. Robinson was beginning a long period of research in the department. Beginning in 1911, he produced 75 scientific publications and bulletins over the next 40 years. A few of these involved acid soils. The majority of papers were about insecticides, fungicides, spray residues, and related subjects.

In addition to his research duties, Robinson was responsible for supervising state fertilizer and agricultural lime laws. About 1921, the fruit and vegetable industries of the state requested a control law governing the sale of insecticides and fungicides. Robinson wrote this law and assisted in its passage by the Oregon Legislature. After its enactment, he supervised its enforcement until 1931 when the State Department of Agriculture was established. At this time, the state agency enforced fertilizer, agricultural lime, and insecticide regulations.

Robinson also did research in the insecticide and fungicide field, including work on a lime-sulfur solution, the preparation and characteristics of various lead and calcium arsenates, and the development of casein spreaders and oil sprays.

In 1925, the future of the apple and pear industry in Oregon was threatened by an embargo on imports by Great Britain because of excessive arsenical residues. Robinson and Henry Hartman of the Department of Horticulture developed a chemical method and suitable equipment for removing the spray residue. The value of this development was later estimated in the millions of dollars. Without it, the apples and pears would have remained embargoed in Britain indefinitely. Robinson's basic method for residue removal was used throughout the United States for years until new insecticides largely replaced the arsenicals.

Other faculty members made valuable contributions during the period. Harry G. Miller did research in mineral metabolism, particularly sodium and potassium. He established the first small-animal colony of white rats on campus and used them in nutrition studies.

In what was then called the Food Products Department, D. E. Bullis began cooperative work in the late 1920's on the canning, freezing, and dehydration of fruits and vegetables. The scientists developed a method for the separation of prunes into quality grades based on specific gravity. Of much greater value to the state was the process they worked out for bleaching cherries for maraschino use. This basic process, still used in all cherry-producing areas of the country, was developed at a time when growers in Oregon were threatened with serious losses because of the lack of a fresh fruit market for their cherries.

In January 1919, J. S. Jones succeeded H. V. Tartar as Agricultural Chemistry Department head. His main interests were in the fields of soils and cereal crops as well as teaching. For many years, he carried on cooperative soil with the Sherman Branch Station research dealing with the depletion of soil fertilizer elements, particularly nitrogen, under continuous grain cropping. During the aforementioned arsenical spray residue trouble in the late 1920's, he established and supervised operation of a control laboratory in the Medford area. He also investigated the composition of forage crops grown in different sections of the state.

Other staff members carrying out valuable work were J. R. Haag, the first department member to devote full time to animal nutrition, and Miles B. Hatch, who worked on food preservation projects. Haag studied the calcium-phosphorus relationship in animal and poultry nutrition and minor element disorders in various sections of the state (that is, the excess or lack of iodine, magnesium, copper, cobalt, molybdenum, and manganese and their effect on nutrition).

Poultry Science

Meanwhile, research was also proceeding in the Department of Poultry Science. The 1913-1914 OAC catalog listed eight courses. In 1916, the first master's thesis,"The influence of different methods of incubation on the development and chemical composition of chicks to the stage of pipping the shell," was submitted by Clara Manerva Nixon. That same year, Miss Nixon was dropped from the Experiment Station staff as a research fellow because of a cutback in funds.

By 1920, five staff members were on the instruction and research staff: James Dryden, A. G. Lunn, F. L. Knowlton, H. E. Cosby, and G. F. Bell. In 1922, the 15th Annual Meeting of the Poultry Science Association was held on campus, July 25-28.

In 1913, the first experiment on poultry feeding was reported in OAC College Bulletin 81. "In an experiment at the Oregon Station, kale painted the yolks a good color of yellow," read the report, which also featured a full page of photographs in color. James Dryden's book, *Poultry Breeding and Management,* was published by Orange Judd Company in 1916 and was widely used as a text.

In 1927, the Poultry Building was dedicated. This three-story brick building, now called Dryden Hall, was also occupied for years by the Department of Veterinary Medicine. With the completion of this building, all OAC chickens were brought to an adjacent central poultry plant.

Agronomy

When the Agronomy Department was established in 1907, it had three divisions: farm crops, farm management, and farm mechanics. Henry Desborough Scudder headed the department from 1907 to 1916. The only other staff member was George R. Hyslop, in charge of the farm crops and farm mechanics divisions. He became head of the

Farm Crops Department when it was formed in 1916 and held that post until his death in 1943. In 1909, a drainage and irrigation division and the Seed Testing Laboratory were added to the Agronomy Department.

Between them, Scudder and Hyslop taught 14 courses during their first two years, as well as managing the college farm and conducting experimental work with soils, cereals, forage, irrigation, and drainage.

Research projects on crops during this period included:

- crop-rotation experiments;
- corn breeding, including early and medium-early strains of the Minn. 13 and Minn. 23 varieties developed by the ear-to-row method;
- breeding and testing of winter and spring varieties of wheat, oats, and barley;
- irrigation experiments on clover, alfalfa, potatoes, corn, mangels, and kale;
- improvements of grasses and legumes;
- breeding and improvement of potatoes;
- trial plantings of hardy rice varieties (none produced seed); and
- resumption of trials on flax and soybean varieties.

Members of the Farm Crops Department analyze hay, 1927. Photo courtesy of OSU Archives, 938.

Station research was already producing these results:

- increased production of clover, vetch, alfalfa, rape, and kale after following Station recommendations;
- the addition of sulfur was found to be essential for alfalfa production in southern Oregon; and
- the foundation was established for the production of improved seed crops, potatoes, wheat, and forage, and for seed certification.

Hyslop devoted a great deal of time to the seed industry in Oregon and started the Oregon Seed Certification Program. He has often been called the father of both the seed industry and the certification program in the state. He first undertook seed certification with potatoes in 1916 to provide seed stock true to variety name and reasonably free of diseases.

In 1917, over one-third of the wheat graded in Portland under Federal standards was classified as "mixed," discounting it as much as 10 cents per bushel less than "pure" grain. In response, in 1918, OAC started the certification of wheat seed stock to solve the mixed grain problem. By 1925, almost no mixed wheat was received from counties active in the grain certification program.

Certification of forage crops seed started in 1924 with the certification of a field of Grimm alfalfa. By the 1960's, the certification program expanded to include 35 crops and over 200 varieties.

Although certification started in Oregon in 1916, the program was not legally vested in any organization until 1957, when a state law officially gave Oregon State Agricultural College responsibility for the program. Certification is now under the direction of extension specialists and an advisory board.

The seed-testing laboratory was started as a USDA cooperative laboratory in 1909 and continued as such until 1954, when it was taken over by the state. The laboratory was part of the Agronomy Department until 1916 and has been a section of the Crop Sciences Department since that time. By analyzing samples and testing for germination, it has been of great service to Oregon's seed industry.

Home Economics

The School of Home Economics faculty with full-time teaching loads and other responsibilities could not spend the long hours of concentrated effort required to go far with lengthy scientific investigations until Congress passed the Purnell Act in 1925. Then, funds became available to support full-time workers in home economics research. Inasmuch as research in home economics was a rather new field, people concerned with it had different ideas of where to place emphasis. "One point clear to us was that home economics research should not try to duplicate what was taking place in other fields," wrote Dean Milam. "Our emphasis needed to be related directly to the home and the responsibilities of the homemaker."[4]

When James T. Jardine, director of the Experiment Station and administrator of the Purnell funds on campus, asked for a proposal, the school had one ready. An intensive study of standards of living in Oregon farm homes in typical communities was proposed. This would provide information needed to improve living conditions and also would be a basis for the future development of home economics at OAC. Jardine was very interested in this line of investigation and allocated funds for a full-time research worker for the project. Maud Wilson, assistant director of the Extension Service at the State College of Washington, was hired for the position.

James T. Jardine was director of the Oregon Agricultural Experiment Station from 1920 to 1931. Photo courtesy of OSU Archives, P46:316.

Maud Wilson had an admirable background for this position, according to Dean Milam. She had boarded in farm homes when she taught in rural schools in Nebraska and had been a state leader for extension work with farm women in Nebraska and Washington. She knew the problems of the rural home. "If the home of today is to be unhampered by the traditions of the past," she once wrote, "there must be a body of knowledge developed to take the place of those traditions which do not meet the needs of the modern homemaker."[5]

Her first study, one of the three projects suggested by the National Committee on Rural Home Studies, dealt with the use of time by farm homemakers. It was carried out in cooperation with the USDA Bureau of Home Economics. Wilson investigated variations in the time-spending patterns in Oregon rural households during weeks of normal activity in relation to: (1) location of home (town or country); (2) major source of income (farm or nonfarm); (3) size and composition of household; (4) schooling of homemaker; (5) household equipment; (6) housework done by persons other than homemaker; and (7) use of commercial goods and services.

The homemakers found Wilson's delineation of their daily activities quite unusual, but they were willing to cooperate. With the help of county agricultural agents, dates were arranged when Wilson could meet with women's clubs and do her research. In general, the response was good, and after a few weeks she became accustomed to competing with ardent quilters, lunch preparations, and other attention-diverting aspects of rural club routine. In all, she reached about 1,200 women in this way.

Wilson made two home visits to each of the women who volunteered to keep the time records. On the first round, she found that some had changed their minds since volunteering, but for the most part she was surprised to find how many were stimulated by the idea and ready to start. She asked each cooperator to keep a time record for a week in which the routine was not broken by unusual circumstances. Each volunteer kept a diary of her own activities and a record of the help she had during the week. The homemakers also added supplementary information on major factors that may have influenced their use of time during the study period.

Wilson obtained usable records in 1926 and 1927 from 513 homemakers—228 in farm homes, 71 from country nonfarm homes, and 154 from noncountry nonfarm homes. Students helped summarize the records. She had a chance to buy a hand-operated relic of an adding machine, built to convert pounds to bushels, and found it a great help in converting minutes to hours. Results of this study were published in various bulletins and periodicals between 1928 and 1932.

Wilson found another outlet for her research. At the University of Chicago, where she later went for graduate study, the data were accepted as the basis for her thesis, "Time Spent in Meal Preparation in Private Households."

The second research project she undertook was called "The Family Home." The study showed that the most important factor affecting the use of homemakers' time was the house itself. She found that building or remodeling was rare in most homes, although some farms were being electrified. In attacking this new problem, she assumed that adequate space and storage was basic to the livability of a house. She found that aesthetics influenced minimum space requirements and she thought her published results would be useful not only to families in building new and remodeling old homes but also to architects and manufacturers. A series of bulletins resulted from this project.

Wilson's plans for farm kitchens had required a preliminary study to set standards for working surface heights and other space units of the house. She queried 312 Oregon and 250 Washington women, who chose preferred heights for working surfaces and chairs.

During the 1930's Wilson collaborated with the USDA Bureau of Home Economics in applying data from its 1934 nationwide survey, about the functions of the farmhouse and its storage requirements. Two USDA bulletins resulted from this assignment.

Food Science and Technology

After World War I, C. I. Lewis of the Department of Horticulture and Dean A. B. Cordley asked Ernest H. Wiegand to come to Oregon to develop courses in food processing and to head a Horticultural Products Section in the Department of Horticulture. Soon after his arrival in 1919, Wiegand instituted the first four-year courses at OAC in horticulture products processing.

From the outset, Wiegand was much in demand all over the country as a consultant to the food processing industry. As early as 1921, he began working to improve the quality of dried fruits, particularly prunes and apples. He began then what has become a long history of service to Oregon's processing industry. Wiegand realized the necessity of working closely with industry to assist with the application of principles and techniques developed through the research program.

In 1925, Wiegand and his assistant, Thomas Onsdorff, developed methods of brining cherries for maraschinos. These methods were widely used, and have been largely responsible for the development of the cherry industry in Oregon, Washington, and other states.

Other early research developments concerned the development of blanching procedures, the improvement of horticultural varieties for processing, the use of simple sugars in the canning and freezing of fruits, and the prevention of darkening by the use of ascorbic acid in frozen foods.

Branch Stations Grow More Slowly

After the rapid growth in the early years of the twentieth century, branch station growth slowed in the next decade. From 1913 to 1931, only three branch stations were opened: John Jacob Astor at Astoria (1913), Pendleton (1927), and the Medford Station (1931).

In 1927, the Oregon Legislature passed an act authorizing an annual appropriation of $2,000 to develop a more profitable system of farming east of the Cascade Mountains. Research projects investigated crop rotation, tillage, varietal testing, breeding and improvement, new crops, soil maintenance and fertility.

In 1928, Umatilla County, the largest wheat-producing county in the state, purchased 160 acres of land suitable for experimental and demonstration work. The County leased the land to the Experiment Station without cost so long as the land was maintained and used as a branch experiment station. An annual Federal appropriation was given in the same year for the study of dryland farming problems in the Columbia River Basin.

The Umatilla County Branch Station farm lay fallow in 1928 and was uniformly cropped to winter wheat in 1929, except for a few acres used for grain varietal trials. Before starting the experimental work, the portion of the field used for crop rotation and tillage experiments was harvested in one-tenth-acre units to obtain information on soil variability. In 1930, the soil of each plot used in the rotation and tillage experiments was analyzed for total nitrogen and organic matter. Similar analysis was done from then on, at intervals of 8 to 10 years, to determine the effect of various cropping systems on the depletion or addition of soil nitrogen and organic matter. The land on the station farm was first plowed in 1891 and the first wheat crop grown on it in 1892. It is probable that no other crop than wheat has been grown on the land since it was first plowed.

Meantime, at the Union Branch Station, researchers once preoccupied with draining, irrigating, and improving valley land for forage and livestock production, now turned to soil fertility work. The Soil Conservation Service conducted an intensive grass adaptation and forage testing program at the Union Station in the early 1920's. Soil fertility work by the Branch Station and the Soils Department on campus during the same period established that deficiencies of sulfur and boron existed. Later, experiments conducted in Baker, Grant, Union, and Wallowa Counties led to the efficient use of fertilizers, mainly nitrogen and sulfur. Experiments in turf grass fields producing seed showed that residue removal was essential to obtaining optimum yields of quality seed.

Staff members at the Umatilla Branch Station continued their testing of fruit varieties. George Waldo, a USDA horticulturist and plant breeder stationed on the Corvallis campus, assisted in testing new selections and breeding crosses of strawberries and raspberries. He found that most varieties of fruits would withstand normal winters.

In the Milton-Freewater area, however, the research program concentrated on production methods, disease, and pest control. Major accomplishments included development of a control program for peach root borer and selection of apple and fruit varieties.

In 1911, when the first experimental work with vegetable crops was established at the Station, curly top vegetable disease was found to be the limiting factor in the production of tomatoes, squash, beets, spinach, and beans. In 1926, M. B. McKay of the Oregon Experiment Station and a research worker at the California Experiment Station simultaneously discovered that the curly top disease was spread by the leafhopper. Concentrated experimental work was started under McKay's direction in 1927 with the active cooperation of the USDA.

At the Sherman Branch Station in Moro, work continued on testing and improving of wheat varieties, and on investigations into cereal breeding, fertilizers, tillage, crop rotation, and crop residues. Several new cereal varieties were developed by branch agronomists: Markton oats, released in 1924; Carleton oats, released in 1938; and Meloy barley, released in 1916 and an improved selection released in 1927.

At the Southern Oregon Branch Station in Talent, research continued during the 1920's to solve problems of the fruit industry, particularly pears. From 1917 to 1919, Station personnel assembled a collection of blight-resistant types of pears and tested them as possible sources of varieties, rootstocks, trunkstocks, or possible material for breeding purposes. The search for such materials was not confined to the United States but ultimately included pears from France, China, and Siberia.

At the Harney Branch Station at Burns, a high altitude area with a short growing season, the research emphasis was on developing improved agricultural practices. At the Hood River Branch Station, the research continued into ways to help the apple industry.

The John Jacob Astor Branch Experiment Station in Astoria ended the decade of the 1920's as it had begun; it continued to study forage and cash crop production and proper dairy management under coastal conditions. In 1929, the Oregon Legislature abolished the boards of regents at the College, the University of Oregon, and the three normal schools. A single State Board of Higher Education was established to administer all the institutions. With control over all budgets, this new entity would come to have an effect on OAC, the Experiment Station, and the rate of agricultural research.

Experimental Highlights, 1916-1930[6]

1916-18 First hen in the world to complete a trap nest record of 1,000 eggs in her lifetime is a white Leghorn bred and reared at Oregon station. War demands for grain bring new livestock feeding experiments. Cross pollination solves cherry orchard sterility at The Dalles. Pear harvesting and storage investigations conducted. Irrigation proved beneficial to vegetable crops in Willamette Valley. Detailed soil surveys under way. Editor of *Better Fruit* says spraying according to station recommendations saved the $1,500,000 apple crop of the Hood River Valley in 1916 which otherwise would have little if any market value.

1918-19 Legislature increases appropriations in 1919, but increased costs and many resignations hamper investigations. Early plowing of summer fallow in eastern Oregon advocated. Important progress in study of Bang's disease reported. Control of liver fluke through destruction of snails started. "Soy beans should become an important crop. Without inoculation they are not successful." A new spring wheat, Hard Federation, obtained from Australia, exceeds other spring wheats at Moro. Fire blight studies in southern Oregon in fifth year.

1920-22 James T. Jardine succeeds A. B. Cordley as director, June 1, 1920. Outstanding results bring increased demand for Station work on many unsolved problems. Both Hard Federation and Federation wheats found suited to practically all of eastern Oregon. New laying record for Barred Rocks established. Recirculating prune driers developed, saving half the fuel, one-third the labor, and producing higher-grade fruit. Oregon could increase her production of hogs to 325,000 or 350,000 per annum, if all waste were utilized. Cost of production studies carried on. Instruments to test maturity of pears. emergency project launched for control of European earwig. Profitable irrigation from wells demonstrated in Harney Valley.

1922-24 Director Jardine reports, "The need of systematic studies in marketing of products of Oregon agriculture is again called to attention. Practically nothing has been attempted in this field by the Experiment Station staff. The problems are of such nature that little will be accomplished without the best effort and full time of at least one well-qualified staff member. The expansion of the station program to this extent is not practicable with the funds and facilities now available." "The many years of experimental work looking to the establishment of alfalfa in the Willamette Valley are beginning to bring success." Various grasses, tried for seed production, continue to give satisfactory yields. Virus diseases of potatoes and bramble fruits studied. Earwig parasites introduced. Forage programs for coast counties developed through use of lime with legumes.

1924-26 More USDA cooperative research established, including resumption of fiber flax investigations. Oregon Committee on Relation of Electricity to Agriculture formed. Federal Purnell Act passed, permitting more attention to marketing investigations. Study of marketing of butter showed that "quality and standardization are altogether lacking." Milk marketing studies pursued. Cause of salmon poisoning in dogs discovered, adding to earlier observations made by E. F. Pernot. Smut-resistant Markton oats introduced. First report presented on studies of coccidiosis in poultry. Experiment Station started research work in home economics.

1926-28 The most outstanding accomplishment of the biennium was the development of a practical solution to the problem of removing spray residue from fruit, an emergency that arose during the winter of 1925-26. It made demands upon all the resources of the Agricultural Experiment Station such as may come only once in half a century. "The value of the station findings has been estimated in millions of dollars. A better measure is the actual fact that the greater part of the fruit crop was practically embargoed until the solution of the problem was found." Western yellow tomato blight found to be in reality curly top, which attacks numerous vegetables. Mosaic found to be the cause of breaking of tulips. New method found to vaccinate for chicken pox. Austrian winter field peas introduced. New greenhouses completed. Irrigated Ladino clover demonstrated as dairy pasture.

1928-30 A process and a machine for separating prunes into quality grades were developed, tested, and covered by public service patents. "This separation will improve the quality of the resulting dried product, speed up drier operation, and not increase the cost of handling." Western Oregon lamb marketing problem investigated, showing possible saving of $140,000 a year. Corvallis strawberry developed. The relation of woolly aphid to perennial apple canker discovered. Walnut blight control partially solved. Irrigation from wells introduced in the Willamette valley. Monthly butter scoring and analysis started. Reseeding of burned-over land studied. Good results obtained from irrigation of small fruits. Methods of bleaching cherries for maraschino developed. Preservation of berries by freezing investigated. Use of crop residues in checking erosion studied at Moro.

Chapter 4 Notes

[1] Quoted in Richard Floyd, unpublished history of the Oregon State University College of Agricultural Sciences, Agricultural Communications, OSU, Corvallis, Oregon, 1983, p. III-3.

[2] Quoted in Floyd, unpublished history of OSU College of Agricultural Sciences, p. III-3.

[3] Quoted in Floyd, unpublished history of OSU College of Agricultural Sciences, p. III-4.

[4] Quoted in unpublished history of the School of Home Economics, Oregon State University, Corvallis, Oregon.

[5] Quoted in unpublished history of the School of Home Economics, Oregon State University, Corvallis, Oregon.

[6] Adapted From "The First Fifty Years of the Oregon Agricultural Experiment Station, 1887-1937," Station Circular 125, August 1937, Oregon State College, Corvallis, Oregon.

CHAPTER 5

The Depression (1930-1941)

The State Board of Higher Education is Founded

In 1929, as noted in Chapter 4, the Oregon Legislature abolished the Board of Regents of Oregon Agricultural College and invested all its powers and duties in a newly established State Board of Higher Education.

At first, the new board seemed willing to continue previous levels of funding for research and extension activities. In the state budget submitted to the legislature by the board for 1931-1933, appropriations supported work at OAC at the same level as before. This was the first time research funding was included in the budget of OAC.

The board's overall budget request for a general unrestricted appropriation totaling $1,181,173 was passed by the legislature without change. Soon after, however, this amount was severely criticized by Governor Julius L. Meier, who vetoed $500,000 of the appropriation and the emergency clause attached to it.

In his letter of March 11, 1931[1] vetoing the $500,000, the governor justified his action by noting that the budget request ran counter to the original millage tax established in 1920 to fund higher education. Under the terms of this act, a state levy of 1.2 mills on a dollar of the total value of all taxable property of the state was made for Oregon State College and the University of Oregon. Four-sevenths of the tax went to the College, three-sevenths to the University.

"When the first millage tax was proposed in 1912," wrote the governor, "assurances were given to the people that if it were enacted into law our higher institutions of learning would seek no further appropriations at the hands of the Legislature. When in 1920 an additional millage tax was proposed . . ., assurances were again given.

"When in 1929 the legislature was asked to create a State Board of Higher Education, assurances were given to the people that, if this board was created, duplications would be eliminated and economies introduced so that there would be brought about a material reduction in the administrative expense of the state's higher institutions of learning. These promises have not been kept. On the contrary, our higher institutions of learning have come to each succeeding session of the legislature and requested additional appropriations. . . ."

The resultant loss of funds was a terrible blow to higher education in Oregon in general and to the research and extension divisions of Oregon Agricultural College in particular. The Experiment Station budget was cut $156,000; the Extension Service lost $87,309. The normal functioning of agricultural research was seriously hampered almost immediately.

What followed was an argument over the original intention of the millage tax. "In recent years, repeated reference has been made by members of the board, at board and committee meetings, which references have been freely quoted in the press, to the effect that the primary function of the board is the teaching of students on several campuses," wrote J. M. Clifford, Extension secretary, in a 1931 report to William A. Schoenfeld, agriculture dean and Extension director.[2] "A direct outgrowth of this policy has been the unwillingness of the Board of Higher Education to sponsor adequate legislative support for research and agricultural Extension programs. . . .

"The campaign which resulted in the passage of this millage tax measure was participated in by friends of higher education in Oregon . . . [who] . . . issued a printed pamphlet of 16 pages which had wide circulation. On page 11 is found a section with the caption 'What the Money Will Be Used For,'" continued Clifford. The last three paragraphs of this section read as follows:

"'6. To develop and promote research. It is the duty of every state to make its share of the annual contribution to the sum of human knowledge, and to the application of new ideas and methods to agriculture and industry. Such a contribution from Oregon cannot be made without proper provision for research.

"'7. To purchase additional lands for building expansion and for agricultural experiment work.'

"'8. To improve and extend the Extension Service of the three schools to the state at large. The training of full-time residence students is by no means all the work of the higher education institutions. . . .'

"It must be borne in mind that the publication was sponsored not only by the College but also by the university and by the Boards of Regents of both institutions. . . ."

This argument would continue all during the 1930's. The state board listed research and extension under a separate category in the budget it submitted to the legislature in 1937. The change was explained as follows:

"The special appropriations listed in this schedule are for the support of certain research, extension, and public service activities which have been requested of the educational institutions by various industrial, agricultural, and social welfare groups and organizations. The State Board of Higher Education, while not listing these appropriations in its regular educational budget, recognizes the value and necessity of the activities and accepts the responsibility for their conduct if the legislature sees fit to appropriate the necessary funds."

In his 1931 report, Extension Secretary Clifford went on to note that the attitude of the board as revealed in this statement seemed to be one in which that body "had no interest whatever in the continuation of these projects, but the facts are quite different. Once having started research and Extension projects, the Board of Higher Education is deeply involved; not only must the board provide rooms, laboratories, heat, light, and janitor service, but it must find and employ trained technical personnel to carry on the work.

"Having induced such personnel to leave previous employment and take up work provided by the special legislative appropriation, the board is inevitably obligated to support the work and sponsor its continuation as long as the board itself is satisfied that the work is making useful progress. It is impossible to get competent help for this type of work unless there is assurance of continuation of the employment. . . ."

Clifford noted the board's moral obligation to support such work and sponsor its continuation. He noted that it was impossible to put a time limit on the successful fulfillment of research. "Such work is carried on in areas beyond the range of human knowledge," he continued. "And as long as the board is satisfied that satisfactory progress is being made toward the solution of the problems for which the appropriation was made in the first place, the board should stand behind such projects with everything that it has and urge the legislature to appropriate the money for continuation of the work, whether or not outside individuals continue to conduct an active campaign for its continuance."

Clifford said that the 1929 law not only put the board in the position of not being interested in the special research projects of the Experiment Station, but absolutely prohibited the College from making any public statement of its vital interests and "even prevents the College from having a fair hearing before the board itself in this connection."

For his part, Clifford proposed that the policy adopted in 1931 be amended by passage of an additional statement:

"Provided, however, that the obligation of the board under the acts of Congress approved March 2, 1887, establishing agricultural experiment stations, and May 8, 1914, establishing cooperative agricultural extension work, which were assented to by the State on February 5, 1889, and February 12, 1914, respectively, shall be considered as coordinate with the teaching of students in the field of agricultural and mechanical arts as set forth on the act of Congress approved July 2, 1862."

Clifford felt that this amendment would cause the board to recognize the equal importance of resident teaching, research, and extension in the field of agriculture as Congress and the Oregon Legislature had originally intended.

The Board Makes Its Presence Felt

By 1932, the new Board of Higher Education had began an extensive rearrangement of people and facilities, under its mandate "to eliminate unnecessary duplication of equipment, courses, departments, schools, summer schools, extension activities, offices, laboratories, and publications."[3]

Oregon State College—the new name agreed to in 1932 but not made official until 1953—became the state's center for all major work deriving from the biological and physical sciences. The University of Oregon became the center for major work in the liberal arts and related fields.

The budget reductions made by Governor Meier and endorsed by the legislature affected more than research and extension budgets. The Board of Higher Education ordered a cut in the salaries of all employees in the State System of Higher Education, effective October 1, 1932.

Before the legislative session, the board had ordered a severe cut in salaries: 5 percent on the first thousand dollars, increasing to 15 percent on the fifth thousand and all amounts above that. Right before the legislature met, the board ordered a second salary cut, ranging from 9 percent on the first thousand to 27 percent on all salaries above $4,000.

The legislature passed a law reducing all salaries of all state employees from 5 to 30 percent, effective March 1, 1933, and continuing until March 1, 1935. By 1934, unemployment had affected more than 10 percent of Oregon's population. Oregon State College suffered accordingly. Nearly 4,000 full-time students had enrolled in 1930. By 1934, there were only 2,245, about one-third of the total registered in the entire state system that year. Between 1930 and 1934, higher education income in Oregon was reduced by 41 percent, primarily by salary and staff reductions. Teaching and research FTE (full-time equivalent, where one FTE equals one position) in the School of Agriculture was reduced by one-fourth.

The Experiment Station faced its greatest challenge—survival.

"The major problem during the past biennium," said a report, "at a time when the economic conditions of the state required a heavy retrenchment of funds, was to maintain the organic structure and the foundation of the technical organization and research built up through continuous efforts over a period of 45 years."[4]

James T. Jardine resigned after 11 years as Station director. In October 1931, the School of Agriculture, the Experiment Station, and the Extension Service were combined under one administrative head; William A. Schoenfeld was named new dean and director.

The 1930-1935 period marked the start of a new era. Everyone learned to make do with less, and be grateful for that, no matter how meager.

In 1937, the Experiment Station, pausing to celebrate its 50th anniversary, named 15 of its top achievements:

1. Discovery of a successful method of removing spray residue from fruit
2. Discovery of the cause and control of Bang's Disease, a destructive dairy cattle problem—leading the nation in eradicating the disease
3. Development of the Oregon small-seed industry through the introduction of new crops, mostly forage, into one of the major enterprises of the state
4. Development of the system of breeding poultry for egg production
5. Development of a practical method of brining cherries for the maraschino trade
6. Production of detailed soil surveys and classification of major farming areas in Oregon, including surveying and mapping about 8.5 million acres of farm land
7. Discovery of a new and simple method of controlling fowl pox, which took a heavy toll in the poultry industry
8. Introduction of important grain varieties, particularly Federation wheat and Markton oats
9. Improvement of Oregon butter quality

10. Development of new methods of storing, ripening, and marketing Oregon pears
11. Discovery of the life history and control of the codling moth under Oregon conditions
12. Control of liver flukes in sheep and goats
13. Development of supplemental irrigation in western Oregon
14. Discovery of the value of sulfur as a fertilizer, first demonstrated in 1912 by the Experiment Station on alfalfa and later shown to be equally valuable for other legumes
15. Control of pear blight, a problem which had eliminated commercial pear production in many parts of the world.

Meanwhile, Back in the Departments

Horticulture

Although horticulture as a subject had been taught at OAC since its founding, and a Department of Botany and Horticulture had been started in 1888, horticultural research at OAC did not achieve national and international prominence until the 1920's.

The leadership of C. I. Lewis as department head (1906-1919) and vice director of the Experiment Station (1913-1919) laid the groundwork for the developments. So did the rapid development of commercial orchards in the state. Most of the faculty of this period and many of its graduate students became recognized leaders in the field. Thirteen of them became department heads at other institutions; three went on to high administrative posts at USDA. Lewis himself became editor of the *American Fruit Grower,* a national trade journal.

This prominence ended, however, with the creation of the Oregon State System of Higher Education and the budget cuts brought on by the Depression. Subsequently, the Department of Landscape Architecture was created, into which all the landscape courses and three faculty members from Horticulture were transferred.

In the interest of administrative efficiency, a Division of Plant Industries encompassing Farm Crops, Horticulture, and Soils was instituted during the Depression, only to be discontinued in 1943.

Poultry Science

In 1930, turkey research started at the Umatilla Field Station at Hermiston in cooperation with the poultry department. The early investigations were concerned with rearing turkeys in semiconfinement and with rations for growing market turkeys. *Cost and Efficiency of Commercial Egg Production in Oregon,* a summary of three years (1925-1928) of cooperative research of the Poultry and Agricultural Economics Departments, was published in 1931.

In 1933, the Poultry Husbandry Department became a part of the newly formed Animal Industries Division of the College. Also included in this division were the Departments of Animal Husbandry, Dairy Husbandry, and Veterinary Medicine.

Farm Crops

In the Department of Farm Crops, research conducted under the auspices of the USDA wheat program developed protocol for crested wheatgrass plantings and grass nurseries, providing the basis for the seeding of more than 200,000 acres of crested wheatgrass in the Columbia Basin within a two-year period.

Animal Husbandry

Professors P. A. Brandt and R. G. Johnson from the Department of Animal Husbandry assisted in the selection of the Squaw Butte Experimental Range site. They were assisted by representatives of the Grazing Service, later to become the Bureau of Land Management of the U.S. Department of Interior in 1934. In 1936, the first range survey was made on the Squaw Butte site. A study was initiated the next year to compare deferred-rotation with season-long grazing on sagebrush-bunchgrass ranges.

Food Science and Technology

In 1938-1939, horticultural products, long part of the Horticulture Department, became a separate Department of Food Products (later to be called Food Science and Technology), where two more Horticulture faculty members went.

In 1939, the Enology Laboratory was set up in the department with H. Y. Yang in charge under an agreement with the Oregon Liquor Control Commission. The objectives of the laboratory were to perform regulatory control work for the commission, to offer consultative service to the alcoholic beverage industry, and to conduct research in enology. The agreement was terminated in 1963.

The Seafoods Laboratory in Astoria was started in 1940 with one staff member, E. W. Harvey. The lab still serves as a center where applied and basic research in marine food science and technology can be conducted in close cooperation with Oregon's fishing industry.

Fisheries and Wildlife

The Department of Fish, Game, and Fur Animal Management was established in 1935. Its name was changed to Fish and Game Management in 1936, and to Fisheries and Wildlife in 1964.

Except during World War II, the department has never lacked students. About 20 were expected in the fall of 1935, but 100 enrolled as majors, even though few jobs were available for professional fish and wildlife scientists during the Depression.

Roland E. Dimick was the only faculty member when the department opened its doors in 1935. He launched the new program with a copy of Aldo Leopold's *Game Management* on his bookshelf and a stuffed pheasant on his desk. Later that year, the importance of the new program was recognized by the U.S. Biological Survey, which founded the Oregon Cooperative Wildlife Research Unit.

In addition to running the department, Dimick also found time to begin a water pollution study on the Willamette River and documented fish kills during 1935 and 1936. Since that time, such studies of the Willamette have continued nearly every year, and departmental research has done much to alleviate pollution. Studies of Oregon's freshwater fish started in 1937. New projects in limnology and trout production were added in 1938. In 1939, the department began to study the native oyster of Yaquina Bay at the request of Oregon Governor Charles H. Sprague. A small laboratory, the forerunner of the Hatfield Marine Science Center, was established for this purpose in 1939.

Pioneer investigations on ringneck pheasants were started in 1939 on the isolated Protection Island near Port Townsend, Washington. Studies on antelope and waterfowl, important parts of the research effort, were directed by Professor Arthur Einarson.

The newly established Oregon Wildlife Federation held its first annual meeting on campus in 1938, a sign that the program at OAC had quickly gained statewide recognition.

Foods and Nutrition

In the early 1930's, the School of Home Economics was able to carry on a little research in foods and nutrition. Agnes Kolshorn devoted a term's work to a study of the baking qualities of three varieties of Oregon-grown pears; two bulletins resulted from this study. Another study on furnishing food requirements at minimum cost resulted in Extension Bulletin 436, "Planning the Family Budget." Findings in a study of the Vitamin B and C content of the Bosc pear and one on basal metabolism were disseminated through the American Home Economics Association.

Nutrition research became well established as an integral part of the Foods and Nutrition Department following the appointment of Margaret Louise Fincke in 1935. Half of her time was allocated to research. A graduate of Mt. Holyoke College, Fincke came to Oregon after eight years at Columbia University, where she had received M.A. and Ph.D. degrees and worked with Henry C. Sherman, an eminent nutritionist who taught several summer sessions at Corvallis. While Jessamine Williams was still head of the Foods and Nutrition Department and after Fincke succeeded her, the department expanded research in nutrition, added foods research as an established entity, and gradually acquired additional staff and laboratory equipment.

Fincke's early investigations included a study to determine the nutritional use of calcium in spinach and kale and studies on the effect of blanching and quick freezing on ascorbic acid, thiamin, and riboflavin in certain Oregon-grown fruits. She investigated the intake of ascorbic acid necessary to maintain tissue saturation in normal adults, ascorbic acid metabolism of college students, and the vitamin value of dehydrated fruits and vegetables. She was a member of an early Food and Nutrition Board of the National Research Council, the group that establishes recommended dietary allowances of essential nutrients for humans.

Andrea Overman Mackey, a graduate of the University of Nebraska, came as an instructor in foods in 1938 and later, after she had returned from completing study for a doctorate at Iowa State College, devoted a major portion of her time to foods research.

In foods research, Mackey studied the quality of frozen meats, the antioxidant effect of edible flours derived from oil press cakes in certain fat-containing food mixtures, and the influence of various production and processing factors on the behavior of fats and oils in food products. In later years, she studied the flavor constituents of cereal grains and the texture of fruits and vegetables and conducted a microscope study of batters and doughs.

While on leave of absence in 1936, Maud Wilson worked with the Rural Resettlement Administration. She visited open-country homes in every section of the United States and cooperated with other home economists interested in the improvement of housing for family living. She helped architects to plan houses for resettled families, and she held group conferences in 10 states. Although she occasionally taught classes, for the most part she devoted her professional career to establishing standards for planning homes to meet the needs of families. Her pioneering work has had far-reaching influence and has provided assistance not only to homemakers but also to designers and builders of equipment for the home.

When Helen E. McCullough came to OSC in 1938 to work with Maud Wilson, she already had a background in housing. She had studied home economics and architecture; she had worked with architects and engineers as a housing consultant. While working with Wilson, she became aware of the urgent need for research in housing, and Wilson provided her with a background of philosophy and experience that later brought her national recognition. After she left Oregon, McCullough worked at Cornell University with Mary Koll Heiner (one-time manager of the campus tearoom). The two women complemented each other—Heiner was a specialist in work simplification and McCullough was a specialist in housing. Later, McCullough continued her work at the University of Illinois and in other positions. She has four books to her credit and has published dozens of bulletins and articles in professional journals. Taken all together, women connected with the OSC staff at one time or another have made a large contribution to home economics research in housing.

Branch Stations Continue to Expand

Despite the Depression and the toll it was taking on the main campus, the branch stations continued to survive or expand.

Meanwhile, at the existing branch stations around the state, research proceeded without interruption though funds were scarce. Somewhat insulated by physical distance and their own dedication to their research, the scientists and other employees of the branch stations worked on—oblivious to the politics swirling around the main campus in Corvallis.

If an industry needed help to survive or expand, the Oregon Legislature and private interests were willing to pay for the research, people, and equipment to do it. Four branch stations were opened during the 1930's.

Medford

The Medford Branch Experiment Station was established in 1931 to study the irrigation, drainage, soil fertility, and related problems of commercial pear orchards in southern Oregon. The Southern Oregon Branch Experiment station at Talent, in the same general district, was already doing scientific pear research. The problem of orchard irrigation, however, was deemed impossible to solve on the limited area of the Talent station.

Through the cooperation of the U.S. Department of Agriculture, Jackson County, the Fruit Growers' League of Jackson County, the Rogue River Traffic Association, and the Oregon Agricultural Experiment Station, facilities were provided and staff assigned to carry out this specialized investigation on a new 20-acre tract, 13 acres of which were in bearing pear trees. The soil of the tract was classified as Meyer clay adobe, typical of 50 percent of the pear orchards of the region.

Squaw Butte

In 1935, the Squaw Butte Experiment Station was established in Burns to investigate ways to improve the range for grazing of cattle and sheep.

Harney

When beef cattle were moved into southeastern Oregon in the 1860's, forage on range lands and marsh meadows was generally free for the taking. Over a period of 50 to 60 years, the vegetation gradually changed from one of sagebrush-bunchgrass to one of mostly sagebrush. Cattle, horses, and sheep were all major factors responsible for the increase of brush and gradual decline of grass. Another factor in the decline of range production was the cultivation of thousands of acres of native range land by homesteaders.

From 1900 to 1920, the U.S. Government gave land to settlers under provisions of the National Homestead Act. As a result, hundreds of people came to Harney County to file on public lands and establish homes on the millions of available acres. In a short time, much of Harney County was completely covered with dry-farm homesteads.

The settlers came with high hopes and all their belongings but, sadly, very little information about crop production and making a living in a semi-arid area (10 to 12 inches of annual precipitation) with an average growing season of 65 days. Many of these settlers had no previous farm experience and had little capital to tide them over the long period required to get established.

By 1911, many homesteaders were in financial trouble. Some had already abandoned their farms and moved elsewhere. The deserted shacks, broken-down machinery, and dust from cultivated areas were mute evidence of their blighted hopes and disappointments. Farmers who remained on their land faced many problems and needed technical help in deciding what crops to grow and how to grow them under the prevailing conditions.

The Harney Branch Station in Burns was established in 1911, "to investigate and demonstrate the conditions under which useful plants may be grown on dry, arid, or non-irrigated lands," according to its charter.[5]

But range livestock production continued to dominate the scene even though thousands of homesteads now occupied the valleys once covered by sagebrush and grass. The beef cattle industry was growing.

In 1934, the Federal Taylor Grazing Act was passed. It brought all public domain lands of the West under controlled management to stabilize the livestock industry and improve the range. It soon became apparent that a great deal of research was needed to accomplish the purposes of the act.

The Oregon Legislature agreed with this general sentiment and, in 1935, enacted legislation "to provide for the purchase of livestock and other capital outlays to be used in establishing a range and livestock experiment station near Burns, Oregon, in cooperation with the United States Department of Interior and to appropriate any money therefore."[6]

Scientists at the new branch station began immediately to work with Harney County ranchers on research to improve native meadows.

Klamath

During the same period, problems of another kind were being experienced by potato growers in the Klamath Basin. In the early 1930's, these farmers were suffering heavy losses to their potato crops caused by the root-knot nematode, which made culls of a large percentage of otherwise fully developed tubers.

The 1937-1939 Oregon Legislature responded to the appeal of potato growers by appropriating $10,000 to start experimental work to control the nematode pest. In 1939, Oregon State College leased from the U.S. Bureau of Reclamation a tract of 80 acres of saline and saline-sodic soil. This tract, representative of about 10,000 acres of such soil in the Klamath Irrigation District, was considered to be too poor for crop production.

The Bureau of Reclamation provided the land, water, and drainage; Klamath County provided funds for buildings and some equipment. The Federal Works Progress Administration cooperated in the construction of a service building and in the installation of 240 rods of woven wire fence.

Red Soils

Soil of another kind formed the basis for establishing yet another branch station. Of the four million total acres of land in the Willamette

Valley, an estimated one and one-half million acres are classified as "Red Hill Soils" by the USDA. Most of these soils are located in Clackamas, Columbia, Marion, Multnomah, Washington, Benton, Linn, Lane, and Polk counties. About half of the farm land in Clackamas County is in this group.

Over the years, much of this land was planted in grain crops, with declining fertility and productivity. Yields on some farms failed to repay the costs of production. Efforts to find substitute crops for grain often met with limited success. As a result of this low productivity, land values declined.

From 1929 to 1939, farm organizations and other groups appealed to the Oregon Legislature to provide funds for research to find methods of rebuilding soil fertility and for testing the adaptability of new crops for the soil. The legislature responded to this need by appropriating $9,720 to establish the Red Soils Branch Station near Oregon City in 1939. Its aim was to investigate the problem of soil fertility drainage, crops, fertilizers, and other problems involving the use and development of the Red Hill lands of the lower Willamette Valley.

The station superintendent was told that this land had been abandoned for farming because of its unproductive condition. When first plowed by station employees in 1940, the ground was covered by a light growth of Douglas-fir trees, Canada bluegrass, and weeds—all indications that the land had not been farmed for several years.

Sherman

The original mission included the search for new cereal varieties of crops that would grow on dry, arid, nonirrigated lands in eastern Oregon.

Wheat is the most important cereal crop in the area, however, and received the greatest attention. Ten varieties were developed and credited to the station in the 1920's and 1930's: Federation and Hard Federation (1920); Hard Federation 31 (1928); Golden (1930); Regal (1926); Sherman and Arco, the first hybridized wheat varieties to be released by the station (1928); Oro and Rio, hard red varieties (1928 and 1931); and Rex (1933).

President Kerr's Chauffeur[7]

G. Burton Wood enrolled at Oregon Agricultural College in September 1929 and majored in commerce.

"I remember the first day I came on campus," he says. "There were orange lines on the sidewalks as soon as you entered the campus property itself. That told you beyond that point not to smoke. There were cans there to deposit your smoking materials. That was my first impression of OAC."

Wood lived in Cauthorn Hall and soon took a part-time job as a student janitor. "I was grateful for the job," he says. "It paid $20 a month, 35 cents an hour. I cleaned the Women's Building from 6 a.m. to 8 a.m. and 5 p.m. to 6 p.m. In my sophomore year I went to work in the Commerce building, where President Kerr had his office. One of my responsibilities was cleaning the president's office every day. I didn't know Kerr. I had to keep the conference table polished and his desk polished. He never left anything on it. It impressed me because he had a private toilet. That was really something."

President William Jasper Kerr was furnished with a car by the state, and it was his practice to hire a student to drive him on trips out of town. He usually walked while he was in town. A white clapboard house was provided for him on Monroe Street (on the site of the present Rogers Hall), so he just walked to his office.

At the end of Wood's sophomore year, Kerr's driver graduated.

"The job paid 50 cents an hour, which would increase my monthly take-home pay," says Wood. "I decided to go to the superintendent of buildings, who was in charge of this job as well as my janitor's post. I was hired and went to work in June 1930. It turned out to be one of the most effective learning experiences of my life."

Kerr's car was a red 1929 Cadillac limousine. Wood kept it polished and serviced in addition to his driving responsibilities.

"I always had to be available and ready to leave within 15 minutes," continues Wood. "He never gave me much time to get ready for a trip. President Kerr was a very formal man. He always called me Mr. Wood."

When summoned by Kerr, usually to drive to Portland or Salem, Wood would back the car up to the front steps of Commerce and then load the car trunk with a number of large volumes and stacks of records Kerr needed for the meetings he was attending. Kerr himself would then appear carrying a small suitcase.

"He always gave me exactly one hour and 50 minutes from the front of commerce to the front of the Imperial Hotel in Portland," he says. "We moved right along up Highway 99. He would take a robe off the robe rail, curl up in the back seat, and sleep the whole way. I would wake him 15 minutes before we arrived."

Wood would spend his time in a room at the hotel while Kerr attended meetings. He took along his school books and studied while he waited to be called to drive Kerr somewhere. Kerr often met with his closest confident, E. C. Sammons, president of the U.S. National Bank and an OAC trustee. Another friend was B. F. Irvine, publisher of the *Oregon Journal.*

"Kerr was a marvelous leader," says Wood. "What impressed me, he could make a decision. If you came to him with a problem, you got an answer before you left his office. He was a skilled administrator who let others take a lot of the credit for things. He had no ego. He was able to bring around him the great deans—Ava Milam in Home Economics, and George Peavy in Forestry, for example."

In 1932, Kerr was asked to become the first chancellor of the Oregon State System of Higher Education. To take the job, Kerr would have to move to the chancellor's office on the campus of the University of Oregon in Eugene.

Continues Wood: "President Kerr said to me, 'You know me, where to go.' He wanted me to be his driver. I was reluctant, but it was like the president of the United States asking you to do something. I couldn't say no."

So Wood transferred to the University of Oregon and continued to work as Kerr's driver. The Cadillac, however, remained at OAC. The vehicle that went with the chancellor's job was a Studebaker, "a real clunker," according to Wood.

In January 1933, Kerr faced the Oregon Legislature for the first time as chancellor. Wood spent weeks holed up at the Marion Hotel in Salem. "I went to class 12 times that first quarter," he says.

When he did go, he often faced problems in class. "There was a lot of rivalry between OAC and the U of O then, and professors would ridicule me in class for being Kerr's driver. I finally had to drop out."

After this, he drove for Kerr for two years.

"I learned so many things from him," he continues. "He'd say to me, 'I know you wonder why I have you load up the trunk with material and not use it. If I needed it, and didn't have it, I'd lose an advantage for Oregon State.' He also told me, 'We make a fuss about A students, but B students run the world.' He was always courteous and never got angry."

When Kerr and Wood arrived in Salem in January 1933 for the first time, the banks had been closed in order to avoid a run on them. The chancellor had not expected that, so he was caught short for probably the first time in his life.

"President Kerr said, 'Mr. Wood, do you have any money? Would you have a $10 bill?' Fortunately, I did have money. 'Could I borrow it from you, I'm out of money.' "

After two years, Wood decided to return to school and reentered the University of Oregon where he graduated in business. He got a master's in agricultural economics at OAC and a doctorate at the University of Wisconsin. After a stint at the War Production Administration, he joined the staff at Purdue. In 1951, he returned to OSC as chairman of the Department of Agricultural Economics. In 1965, he became director of the Oregon Agricultural Experiment Station until 1975.

The Agricultural Research Foundation Eases the Depression[8]

The hard times at OSC's School of Agriculture and Agricultural Experiment Station were eased somewhat by the establishment of a new organization, the Agricultural Research Foundation, designed to facilitate and encourage research in all branches of agriculture. From its modest start in 1934, with the first grant of $1,000 to study the use of sulfur in agriculture, to the more than 400 projects and assets of over $2 million in 1984 at the time of its 50th anniversary, the foundation has played an important role in Oregon agriculture.

Two men were responsible for getting the foundation started. William A. Schoenfeld, dean and director of the School of Agriculture, and Robert M. Kerr, a Portland attorney and son of former OSC president, William Jasper Kerr, first met to discuss the idea in 1934.

Schoenfeld brought with him an article about a foundation at the University of Wisconsin that handled royalties from a university project. He thought it was time for Oregon State College to have something similar. Kerr agreed, and the dean had soon recruited two other charter members, Judge Guy Boyington, an Astoria civic leader, and R. L. Clark, a Portland business, civic, and agricultural leader.

The nonprofit, charitable, scientific, and educational corporation was organized on October 27, 1934, two days after the articles of incorporation were signed. The purpose of the new foundation was to stimulate and provide funds for agricultural research, particularly when state funds were so tied to regulations that they often delayed research projects and grants.

The articles also called for the foundation "to solicit and receive donations, gifts, scientific works and materials, letters, patent applications, copyrights, trade-marks and trade names, both foreign and domestic." Its sources of revenue and income were to be gifts, grants, and voluntary contributions, as well as income from the use of "inventions, scientific discoveries, patents, trade-marks, copyrights, scientific formulas, and other similar properties. . . ."

The three trustees decided the new foundation should have 11 directors, each to serve three years. One would be the dean of the School of Agriculture, one the director, and one the associate or assistant director of the Oregon Agricultural Experiment Station. Eight were to represent the major agricultural interests of Oregon. Two would be members at large.

All members of the new corporation attended the first meeting of the foundation on March 22, 1935, at the Imperial Hotel in Portland. Schoenfeld was temporary chairman, and Ralph S. Besse, vice director of the Agricultural Experiment Station, was named temporary secretary.

The first matter considered was a proposed contribution by Texas Gulf Sulphur Company to finance an Experiment Station investigation of the use of sulfur in agriculture. After discussion, the matter was referred to Kerr, the foundation's attorney, who prepared the forms of agreement.

At later meetings, the trustees agreed to accept a $6,000 donation from the Wild Life Institute for a research project on fish, fur, and wild animal management. They also approved a study of coccidiosis-immunity in poultry, an investigation of apple and pear by-products and utilization, and an investigation of the use of sulfur in agriculture.

The partnership between the College of Agriculture and leaders of agriculture was firmly established from that first meeting. With only minor modifications in meeting patterns and operating methods, the basic thrust of the foundation would continue for the next 50 years and beyond.

The projects funded by the foundation in the years since represent a veritable history of agricultural research in Oregon:

- Investigating the pea weevil
- Studying leaf arsenate spray
- Examining the effect of corn sugar on freezing of small fruits
- Researching the use of potash in agriculture
- Researching brined cherry pack of The Dalles Cooperative Growers
- Studying wildlife and game problems in Oregon
- Sugar beet seed development, maturity, and production under Oregon's climatic conditions
- Looking for new and improved methods of using fruit pulp
- Studying game herds on an island under natural conditions
- Investigating to determine the accuracy of certain methods and procedures in sampling, preserving, and testing of milk received at milk plants

- Looking at the effects of hormone sprays on the canning quality of Bartlett pears
- Investigating field production of pyrethrum
- Studying the use of copper in agriculture
- Studying day-drying
- Researching field and seed crops, and forage on alkali soils and peat soils at the Klamath Branch Experiment Station
- Gathering, assembling, and interpreting data relating settlement experience and credit requirements of settlers on undeveloped reclamation farms for the U.S. Bureau of Reclamation
- Developing improved methods of field distillation of peppermint oil
- Developing economic injury levels and information for aphids affecting Oregon wheat
- Studying the potential for soybean production in Oregon
- Developing the field of clinical immunology and its application to the management of neonatal infection in calves
- Studying the prevention of retained placentas in dairy cattle
- Examining diet composition, food interfacing, and predicted stacking rates of grazing cattle in eastern Oregon
- Looking for mechanisms of resistance in wheat cultivars
- Examining the response of forage to grazing systems on foothill rangeland
- Studying possible genetic mechanisms of chalk brood disease resistance in the leaf-cutting bee.

Selling the results and benefits of agricultural research became a priority program for the foundation. In the belief that there is a keen interest in scientific research, the foundation provided support for making research findings available to the general public via television, radio, and magazines. The "Magic of Research" slide set, presented by R. W. Henderson, became a well-known story about agricultural research and was shown to many audiences throughout the state.

It is the Foundation's philosophy that agricultural support will flourish and grow as the results of the research are translated into benefits that the general public can recognize and appreciate.

Experimental Highlights, 1930-1941[9]

1930-32 No complete report of the Experiment Station published for this biennium. "The major problem of the Experiment Station during the past biennium at a time when the economic conditions of the state required a heavy retrenchment of funds was to maintain the organic structure and the foundation of the technical organization and research built up through continuous efforts over a period of 45 years." James T. Jardine resigned after 11 years' service. Dean A. B. Cordley became dean emeritus, and the School of Agriculture and the Agricultural Experiment Station and Extension Service were combined under one administrative head in October, 1931, with Wm. A. Schoenfeld as dean and director. A total of 391 problems studied during the biennium. Cherry fruit fly controlled. Cooperative small-fruit breeding started with the USDA. Ripening methods for Bosc pears discovered. More than 30,000 acres in the Willamette Valley are now devoted to alfalfa.

1932-34 Crested wheat grass, introduced at the Moro Branch Experiment Station some 10 years ago, becomes most promising forage grass for eastern Oregon. Rex wheat, a smut-resistant, hardy hybrid developed. Parasite of woolly aphid, introduced at Hood River Branch Station, shows effect in control. Chemically treated fruit wraps prevent scalds and rots in storage. Electricity applied to soil sterilization, hotbed heating, corn and hop drying, and home-made brooders. Relation of moisture and irrigation to pear production in Rogue River Valley studies at Medford Branch Station. Grass established in waterways in wheat fields to check erosion at Pendleton Branch Station. Baby beef fattening proved practical in eastern Oregon.

1934-36 Research units in wildlife conservation established in cooperation with Bureau of Biological Survey and State Game Commission. Use of acidophilus milk found to be cure of dysentery in young lambs and calves. Ethylene gas, discovered emanating from ripe pears, found to explain a number of baffling storage problems. Emergency attention of entomologists directed against spittle bugs and pea weevils; distinct progress made in control. Report on public expenditures in Oregon as measured by property tax levies. Range livestock station at Squaw Butte consisting of 16,000 acres of range land established to serve entire West.

1940-41 Alta tall fescue was introduced; it became one of the leading forage and soil conservation varieties grown in the United States. Serious virus diseases affected cherries in The Dalles and led to the establishment of The Dalles Experimental Area. John A. Milbrath and Sam M. Zeller identified a number of diseases caused by the virus.

Chapter 5 Notes

[1] Internal memos, Oregon State University Extension, on file, Director's office, Agricultural Experiment Station, OSU, Corvallis, Oregon.

[2] Internal memos, OSU Extension.

[3] Quoted in Richard Floyd, unpublished history of the Oregon State University College of Agricultural Science, Agricultural Communications, OSU, Corvallis, Oregon, 1983, p. III-5.

[4] Quoted in Floyd, P. III-6.

[5] Unpublished history of Harney Branch Experiment Station, Burns, Oregon.

[6] Unpublished history of Harney Branch Experiment Station, Burns, Oregon.

[7] Based on Ron Lovell interview with G. Burton Wood, March 8, 1988.

[8] *Agricultural Research Foundation 1934-1984,* Oregon State University, 1984, p.3.

[9] Based on "The First Fifty Years of the Oregon Agricultural Experiment Station, 1887-1937," Station Circular 129, Oregon State College, Corvallis, Oregon, August 1937.

CHAPTER 6

World War II (1941-1945)

OAC Goes to War—Again

As the ravages of the Depression eased, the Agricultural Experiment Station and its parent institution, Oregon Agricultural College, faced an even greater catastrophe: another world war. Some of the same events of 1915 happened again. Enrollments in the School of Agriculture fell. Many faculty members left campus for war work. Military training became a responsibility of the College.

OAC became the site of the West's first Army Specialized Training Program. Launched in 1943, this program provided troop technical training in engineering, communications, and languages. Two thousand men and women were trained on campus.

Before long, about 9,000 graduates and former students of Oregon Agricultural College were in uniform. Emergency projects in war programs took priority in the schedules of Experiment Station researchers and Extension Service personnel.

As they had done in the Depression, extension agents took the lead in establishing county emergency programs. They prioritized farmers' needs for equipment, building materials, and repair materials. Agents organized a salvage program and worked with farm labor committees on distributing farm labor. They shared their expertise in publicity and organization with the Food for Victory and Victory Garden programs.

A telegram sent by Governor Charles Sprague to Chancellor F. M. Hunter reflects the general acceptance of the Colleges' ability to deliver:[1]

> A critical situation faces poultry brooders through shortages of fuel. I wish to assign to College Extension Service task of surveying fuel needs through county agents. Kindly direct Extension Division to assume this duty immediately.

The work was immediately done.

As often happens anywhere in periods of uncertainty, the campus was often plagued by rumors. In February 1942, for example, George R. Hyslop, head of plant industry on campus, wrote George W. Peavy, dean of Forestry and coordinator of war activities on campus, worried about a possible problem with so many military personnel nearby.

Peavy replied:[2]

> It was to be expected that questions would be raised in the minds of many of the friends of Oregon State relative to relation of the College to the cantonment soon to be established north of the city. In spite of all we can do, criticisms will be made and probably some who would otherwise send their girls here will not do so.

While the College was busy training men and women in uniform, its scientists worked in field and laboratory synthesizing antimalarial drugs and on other projects, some of them classified. Another responsibility of both scientists and extension agents was to make sure that an ample food supply was produced, preserved, and distributed.

Oregon Agricultural College women students operated harvesting equipment during World War II in Umatilla County. Photo courtesy of OSU Archives, 972.

The Departments During The War

Some faculty members joined in active military duty during World War II. In the Department of Agricultural Chemistry, for example, Paul Weswig joined the staff in September 1941 as assistant biochemist in animal nutrition. In the spring of 1942, he was granted a leave of absence for military duty. He served as nutrition officer for the Eighth Air Force and in 1946, after he was discharged, returned to the department. He went on to a distinguished career researching vitamin A in dairy products and vitamin A metabolism and requirements in cattle.

Virgil Freed joined the Farm Crops Department in 1945, specializing in the chemistry of herbicides. He divided his service between that department and Agricultural Chemistry until 1954, when he joined the latter full-time. He would eventually become a recognized authority on the chemistry of pesticides, particularly herbicides.

In 1940, the first broiler-related research from the Station resulted in Station Bulletin 386, *Surplus Leghorn Cockerels as Broilers*. The authors were Wilbur Cooney and F. E. Price, both of whom would later become deans of Agriculture at OAC. In the Department of Poultry Husbandry, the Oregon Legislature appropriated $20,000 in 1945 to establish a turkey experimental unit to construct buildings and purchase equipment for conducting research on housing, feeding, and management of turkeys.

The Department of Crop Science and OSC and the Experiment Station suffered a great loss in 1943 with the death of George Hyslop. He died unexpectedly while on a trip to Klamath County to inspect potatoes for certification. Hyslop had just returned from an exhausting trip to eastern Oregon and was tired. To a colleague who pleaded with him to stay home and rest a day or two, Hyslop said: "There are a thousand acres of potatoes to be inspected, and there is no one else to do them."[3]

Hyslop had been instrumental in establishing a crops research facility for OAC in 1928. Hyslop Farm, located about seven miles northeast of Corvallis, has proven to be a valuable research facility. The farm consists largely of Willamette soil as well as sandy, or bottom land soil, both types crucial to the understanding of how to best grow crops in the Willamette Valley, one of the largest agricultural areas of the state.

But Hyslop was responsible for much more. He was the major influence in the formation of the Oregon Wheat Growers League and the Oregon Seed Growers League. He was responsible for the first Oregon seed law and, more than anyone, he foresaw the development of the Oregon seed industry.

At this particular time, range management research was centered in both farm crops and animal husbandry. This research had begun in 1935 with the founding of the Squaw Butte Experimental Range near Burns.

The Branch Stations During the War

The Oregon Agricultural Experiment Station entered the war years with 13 branch stations. Neither budget cuts nor changed priorities brought on by World War II kept the staff members at the branches all over the state from their work.

Union

In 1941, the Hall Ranch, a 2,000-acre tract of foothill range near Union, was purchased to supply summer grazing for the Eastern Oregon Experiment Station. This tract also made concentrated research possible on livestock-range-forestry problems encountered on the privately owned foothills of northeast Oregon.

Hood River

War or no war, scientists on the main campus and at the Hood River Branch Station had to turn their attention to virus diseases affecting stone cherries in The Dalles area in 1941. J. A. Milbrath and S. M. Zeller identified a number of diseases later found to be caused by the same virus.

By 1943, growers were deeply concerned with the spreading stone cherry disease, and they appealed to the Oregon Legislature for research funds to help halt the malady. The legislature responded with an appropriation of $5,000 for a two-year investigation of "disease, pest, nutritional, cultural, and soil problems of stone fruits." The new financial assistance enabled OSC scientists to do more research in The Dalles area.

Although the research was intensified, the disease spread. A survey by the Oregon Department of Agriculture showed 3.7 percent of the cherry trees inspected to be affected with the so-called "little cherry virus." In some orchards, the infection appeared in more than 50 percent of trees. The cherry industry appeared to be doomed. The growers appealed to the legislature for still more research funding. The 1947 legislature appropriated $20,000 for this purpose, but the virus eluded cure.

Malheur

In 1942, farmers in Malheur County asked Oregon State College to establish a branch station in their area. They raised money to buy land for the station, and this land was eventually deeded to OSC in the 1950's.

The Malheur Branch Station at Ontario specialized in research important to alfalfa hay and such new crops as sugar beets, onions, and potatoes. Since its inception, the Station has conducted variety evaluations of onions and studies to combat thrips, weeds, and *Botrytus* neck rot. Artificial drying studies have evaluated the use of heat to kill or inhibit neck rot fungi.

Potato variety evaluations, irrigation studies, and soil management research at the Station emphasized minimizing the sugar-end problem. Scientists sought cultural practices minimizing water and heat stress on potato plants.

Sugar beet research included work on mildew, seed emergence, variety evaluations, weed control, and irrigation. Weed-control research was aimed at finding herbicides or herbicide combinations that curb weeds without injuring crops.

Facing the Future

As the war ended in 1945, both OAC and the Oregon Agricultural Experiment Station faced the future with optimism. The Station and its scientists would be more needed than ever to feed people in the nation and the world.

The College would be faced with burgeoning enrollments. August L. Strand, who had become OAC president in 1942, was worried about a problem that had plagued such past presidents as Benjamin Arnold and William Jasper Kerr: the need for balance between education and training.

Strand addressed the State Board of Higher Education on this subject during the war:

> The student who . . . puts a liberal goal first should nevertheless emerge with a means of livelihood; the student who puts his professional preparation foremost should emerge with at least the essentials of a liberal education. Oregon State College must steer a course directly between a futile liberation on the one hand and an excessive vocationalism on the other.

Experimental Highlights, 1941-1945

1940 Alta tall fescue, which became one of the leading forage and soil conservation varieties grown in the United States, was introduced.

1941 Serious virus diseases affected cherries in The Dalles and led to the establishment of The Dalles Experimental Area. John A. Milbrath and Sam M. Zeller identified a number of diseases caused by the virus.

1942 Malheur Branch Station was established near Ontario to work on irrigation and soil management problems in the Snake River Valley.

1945 A turkey experimental unit was established near Corvallis for research on housing, feeding, and management of turkeys.

Chapter 6 Notes

[1] Quoted in Richard Floyd, unpublished history of the Oregon State University College of Agricultural Science, Agricultural Communications, OSU, Corvallis, Oregon, 1983, p. III-9.

[2]Quoted in Floyd, P. III-9.

[3] Unpublished history of the Department of Crop Science ("The Evolution of the Crop Science Department, 1916-1959," by Don Hill), Corvallis, Oregon, 1985.

[4] James W. Groshong. *The Making of a University, 1868-1968.* Corvallis, Oregon State University, 1968.

CHAPTER 7

The Postwar Era (1945-1965)

The Oregon Agricultural Experiment Station Rebuilds

The years immediately after World War II were ones of rebuilding for Oregon Agricultural College and its Agricultural Experiment Station. The rebuilding was both on paper and in fact—on paper because of the severe budget cuts brought on by the reorganization and the Depression; in fact because some programs, buildings, other facilities, and even fences had been neglected while more urgent war tasks had been addressed.

The lines of responsibility for both Federal and state programs had long been established. The Experiment Station had established a network of stations across the state. Scientists working at these sites and on the main campus in Corvallis were ready to work.

Horticulture

E. J. Kraus, Department of Horticulture, achieved early prominence as a scientist after the war. His Ph.D. dissertation, started at Corvallis, and that of H. R. Krayhill of Pennsylvania State College, were published jointly as Station Bulletin 149. "Vegetation and Reproduction, With Special Reference to the Tomato" relates the internal biochemistry to growth and reproduction. It became the foundation of much of the basic and applied research in plant physiology for many years.

Henry Hartman was another horticulturist who won national recognition, specifically for his research on storage and handling of winter pears. He developed the "Hartman Wrap," a chemically impregnated paper wrap widely used on nearly all Anjou pears grown in Washington and Oregon. Elmer Hansen, named assistant in horticultural research in 1935, also gained national and international recognition for his work on postharvest physiology of fruits.

The department itself experienced a period of growth after World War II that resembled the phenomenal one of the 1920's in many respects. Returning veterans and an expanding agricultural industry assured increased enrollments and friends for research and extension. The vegetable crops staff was expanded to take care of the growing research needs of the vegetable industry.

A great deal of research during the postwar period was done on the influence of fluoride emissions from Columbia River aluminum plants on orchards and other crops in that area.

Agricultural Chemistry

In 1946, J. S. Butts took over as head of the Department of Agricultural Chemistry. He was particularly interested in the use of radioisotopes as tracers in agricultural chemical research. Through his efforts, the Atomic Energy Commission made the first Federal grant to OAC in 1951; the grant was continuous until 1961. He also did extensive work in the field of carbohydrate and amino acid metabolism in human nutrition.

Poultry Science

Research on turkey reproduction problems began in 1946 with the installation of 160 young hens and 20 toms in two of the chicken houses at the South Farm. In the spring of 1947, turkeys were reared on the Wyatt Farm and on the Turkey Farm by the next fall. A grant of $20,000 from Swift and Company, to be spent over a five-year period, greatly aided this research. From 1954 to 1959, nutrition and physiology laboratories were set up in the main poultry building.

Farm Crops

The Farm Crops Department, as well as OAC as a whole, experienced a surge in enrollment after World War II. It was necessary to increase staff at a time of intense competition for qualified people.

Growth also occurred in one department program, the Seed Laboratory, first set up as a USDA cooperative facility in 1909. Originally, the USDA was interested in the regulatory samples required by the Seed Act. As the seed industry grew in Oregon, however, the percentage of time spent analyzing the Federal regulatory samples became less and less. By the early 1940's, the laboratory had become a wholly local operation. A fee system was established, and, except for some facilities provided by the College, the laboratory was self-supporting.

During the war years, Oregon farmers had been asked to grow seed crops as their part of the war effort. Experts from other places finally became aware of what Hyslop had known for years: the western Oregon climate was ideally adapted to the production of many types of agricultural seeds. The government contracted to buy the seed at a specified price. The contract also stipulated that the seed had to be tested.

Don Hill, head of the Department of Farm Crops from 1943 to 1959, describes department facilities during those early days: "By now, the laboratory had outgrown the one small room it had occupied for years and was using one end of the resident instruction laboratory. The facilities were utterly inadequate to meet the demands that were put upon it in the fall of 1943. By November, the laboratory was about four months behind in its work. Farmers could not be paid for their seed crop until the tests were available. Every time the telephone rang it was usually some irate farmer asking about his test. The situation was intolerable. When the Seed League met about December 1, I requested to be allowed to speak to the convention as first speaker."[1]

Hill asked for better facilities, higher fees, higher wages for lab workers, and the establishment of a research program. The league appointed a committee to deal with the problem and requested a meeting with the president of OAC and the chancellor by the following Monday. Soon thereafter, all of his requests were granted.

Range Management

Meanwhile, in Range Management, a separate curriculum was being set up. In 1950, the first Bachelor of Science degrees in that field were awarded, although the program was still part of crop science. That same year saw the start of a range ecology project in eastern Oregon that in 1954 became a tristate regional project with Washington and Idaho.

In 1951, E. R. Jackman began developing 137 alfalfa-variety nurseries that eventually provided background information for extensive seeding of grazing-type alfalfa throughout the Pacific Northwest. In 1953, Jackman joined the Oregon Extension Service as its first range management specialist.

In 1961, a cooperative effort with the Research Division of the Oregon Game Commission was undertaken with the start of projects on "Evaluation of Game-Range Habitat." That same year the first Bureau of Land Management-supported research project on "the control of medusahead on Oregon ranges" began.

Fisheries and Wildlife

In 1941, Fisheries and Wildlife established a fisheries curriculum. The department began conducting pollution studies on the Willamette River in 1944; in 1945, the U.S. Fish and Wildlife Service formed the headquarters for the Western Fish Culture Investigations under a cooperative agreement with the department and the College.

By 1948, department research was attracting national attention. Arthur Einarson received two national awards for his book, *The Pronghorn Antelope and Its Management.* The National Council for Stream Improvement entered a cooperative agreement with the OSC Engineering Experiment Station for conducting studies on pulp mill wastes; the Fisheries Department was designated to conduct the fishery aspects of the research.

In 1953, the U.S. Public Health Service transferred a Toxicity and Fishery Unit from Cincinnati, and placed it in the department as the Pacific Cooperative Pollution and Fisheries Research Laboratory, under the joint direction of Peter Doudoroff and Charles E. Warren, bringing national and international recognition to the department.

The Oak Creek Laboratory of Biology and the Berry Creek Experimental Stream were used to train students in water pollution. By 1957, this work had attracted an additional five-year grant to investigate the resistance of marine organisms to various pollutants. The efforts at Oak Creek also had a great influence on the later decision to place the U.S. Environmental Protection Agency Regional Laboratory on campus and to locate the Western Fish Toxicology Station in Corvallis.

Animal Science

Animal Science was a new department, set up after the war. Although one of the oldest (1907) divisions of investigation of the Oregon Agricultural Experiment Station dealt with domestic animals, a specific research program in animal science did not begin until 1947.

Early departmental emphasis was on dairy cattle and dairying. Eventually, animal studies were expanded to include beef cattle, draft and riding horses, along with laboratory animals, mink, pigs, and sheep. The approach was at first mainly one of demonstration: showing what performance was possible with the different livestock species, managed according to the best available methodology.

More recently, the department conducted controlled scientific investigation. A significant early achievement, cosponsored with Veterinary Medicine, was the successful control of Bang's disease, or infectious abortion of cattle.

The transition from animal demonstration to research began with the appointment of Fred McKenzie as head of the Department of Animal Husbandry in 1947. McKenzie, a distinguished animal physiologist trained at the University of Missouri, brought enthusiasm for investigation and dedication to scientific methods that were soon picked up and applied by his staff.

The department was merged with dairy husbandry and food technology in 1954 and did not become the Department of Animal Science until 1959. As financial support allowed, research was done on animal breeding, genetics, feeding, nutrition, and reproductive physiology with all species of domestic animals available.

In the mid-1950's, Dean and Director Earl Price asked the department to investigate a troublesome problem called "white muscle disease" which annually killed thousands of young calves and lambs in the rangelands of central Oregon's volcanic plateau. A cooperative project was established involving LeMar Remmert (Agricultural Chemistry), O. H. Muth (Veterinary Medicine), and James Oldfield (Animal Science). The disease was reproduced experimentally by bringing to Corvallis feeds from affected areas. In 1959, the research team discovered the cause of the problem: a deficiency of a little-known trace element, selenium. They found that white muscle disease could be completely prevented or cured by giving very small amounts of selenium by mouth or injection.

Beyond its obvious usefulness to Oregon ranchers and consumers of their products, the selenium finding led to Oregon's recognition as a center for trace-element research.

Food Science and Technology

Originally a part of the Horticulture Department, the Department of Food Science and Technology acquired its present name in 1962.

Beginning in 1917, the department conducted research on food chemistry, food toxicology, and food processing. The College became one of the first three institutions in the United States to offer courses of this nature.

In 1949, the department researched diets for hatchery-raised salmon and trout. Scrap fish and waste from the fillet lines at the salmon and tuna canneries were tested as fish food but were found to spread diseases such as salmon tuberculosis to the hatchery fish.

OSU Food Science and Technology researchers eventually developed a process to pasteurize fish cannery scraps and virtually eliminated the transmission of salmon tuberculosis and other diseases spread when offal and other fish scraps were used for food.

In 1954, feeding experiments were conducted to test the production-type diets. Moist pellet were developed and kept frozen until feeding. From 1955 to 1958, this Oregon Pellet was tested in large-scale breeding trials at the Oregon Fish Commission hatcheries. In 1959, commercial production of the pellet began at Warrenton, Oregon, and it was used at the 16 commission hatcheries. The culmination of almost 14 years of research resulted in the largest total salmon and steelhead take ever in 1963.

Agricultural and Resource Economics

As with a number of other station-sponsored departments, Agricultural and Resource Economics came into being a lot earlier (1917) than it achieved departmental status under its present name (1949). It began life as the Farm Management Division of the Department of Agronomy. A separate "agricultural economics" section in the School of Commerce was developing at the same time.

After the Oregon State Board of Higher Education dissolved the School of Commerce in 1933, agricultural economics courses were taught by the Department of Economics in a new service school known as Lower Division. Some of the faculty held joint appointments in the Oregon Agricultural Experiment Station, conducting research on agricultural marketing, transportation rates, and taxation. Farm Management continued on a separate course in the School of Agriculture.

In July 1949, the Department of Farm Management was discontinued and its courses transferred to a Department of Agricultural Economics along with the agricultural economics courses taught in the Department of Economics.

The department took on its present orientation in 1951 when G. Burton Wood was appointed head, a position he held until he became director of the Oregon Agricultural Experiment Station in 1966. The primary clientele served by the department in its early years were those individuals and groups involved in the production and distribution of agricultural products. Research emphasized efficiency of production, marketing, and farm income and price policies.

School of Home Economics

In the 1940's Maud Wilson from the School of Home Economics collaborated with Professor H. E. Sinnard, professor of architecture, in developing a series of sketch plans for rural Oregon houses. The results were published in 1945 as a 209-page handbook for use by extension workers, *Plans for Oregon Farm and Acreage Homes.*

In 1947 Oregon cooperated with other states of the western region in a field study of the housing needs of western farm families, designed to set geographic boundaries to farmhouse design.

Clara A. Storvick, with degrees from St. Olaf College, Iowa State College, and Cornell University and with research experience at Oklahoma Agricultural and Mechanical College, Cornell, and the University of Washington, joined the nutrition research staff in 1945.

In addition to her research, Storvick was also a skilled administrator. In 1952 she received the Borden Award of the American Home Economics Association for leadership in regional research on nutritional status. In 1955 she became the first to chair home economics research in the Agricultural Experiment Station; 10 years later, when a

Nutrition Research Institute was formed to coordinate the work of various scientists on campus concerned with nutrition, Storvick was named its director.

Helen G. Charley, a graduate of De Pauw University and University of Chicago, joined the foods staff in 1944. Charley experimented with the effects of the size, shape, heat penetration qualities, and construction material of pans on the quality of baked products. She investigated the chemical composition of fats and how they are affected by cooking at various temperatures, and she provided data to determine a precise method of cooking salmon. She also studied various pigments found in fruits and vegetables and their effect on color, flavor, and keeping qualities.

A School of Home Economics nutrition research project on dental caries had wide-reaching results. Physical examinations of Oregon men in military service in World War II revealed what appeared to be an excessive amount of tooth decay. The Extension Women's Council helped persuade the Oregon Legislature to provide funds for an investigation of the effect of nutrition on tooth decay. Initiated by Storvick and Demetrios M. Hadjimarkos and later carried on by Gertrude Tank, this investigation provided basic data that have helped in understanding the relationship of diets and dental caries. Research to promote dental health continues today with F. Cerklewski's research on fluoride bioavailability and the influence of maternal alcohol intake on dental health of offspring, using an animal model.

The experiment stations of the Western states expanded the dental caries investigation in the late 1940's to include samplings in Idaho and Washington. In other regional studies, the nutritional status of about 2,000 teen-age boys and girls in nine Western states were examined. Dietary intake was correlated with blood nutrients. Another phase of the project, confined largely to California and Colorado, studied the nutritional status of aging people as influenced by age, sex, and food habits. More than 60 papers based on this regional research, about one-quarter of them with Oregon scientists as authors or coauthors, were published in journals and bulletins between 1951 and 1958.

In textiles research, the Department of Clothing, Textiles, and Related Arts took an active part in a linen-weaving project. Flax grows well in the Willamette Valley, and it produces linen of good quality. To reduce the national dependence on overseas sources and to encourage the development of a new agriculture-based industry in the state, various Federal, state, and private agencies supported extensive flax production and linen-utilization studies.

The department designed and tested fabrics using Oregon linen. The creative work of Joan Patterson showed that the fiber could be made into a variety of beautiful, colorful, durable textiles. Household fabrics made of Oregon yarns compared favorably with those made from imported yarns. Linen weaving, however, did not grow into a big industry for two reasons: the high cost of turning fibers of the flax plant into linen yarn, and the low cost of synthetic fibers coming onto the market, replacing linen in household use.

Other clothing and textile research projects included Clara Edaburn's research on the design and construction of functional house dresses for the mature figure, Ida Ingall's project on changes needed in current ready-to-wear school dresses for the 7-14 age group, Florence E. Petzel's work on the thermal properties of blankets of different fiber content. The first Western Regional Project in Textiles was launched when Oregon joined other states in studying the effects of atmospheric conditions on specific cotton fabrics. Another regional project studied the efficiency and cost of laundering textiles with different detergent types and water temperatures. Other important projects in clothing and textile research included comfort factors in clothing, the nature of natural fibers, and the need for consumer protection.

In the Department of Household Administration, which later became the Department of Family Life and Home Administration, Katherine Haskell Read (Baker) came as supervisor of the nursery school in 1941 and after Sara Prentiss retired, became department head. Read carried on many investigations in child growth and development and, in books and articles she wrote, pointed out what an excellent human relations laboratory the nursery school makes. Lester Kirkendall, the first man on the teaching staff of the School of Home Economics, joined the department in 1949. The research for which he became widely known across the country, and to some extent overseas, centered around his quest for a value framework to describe the importance of interpersonal relations.

In her book, *Adventures of a Home Economist,* Dean Milam concluded:[2]

> "One of the most gratifying features of the home economics research carried on through the years on our campus has been the scientific spirit of the whole staff. They were not content to base their teachings on the worn-out or grown-out theories of the past; they continually probed for better ways of doing things, better theories on which to base decisions, and new methods to apply to the ever-changing conditions of the American home. They have benefitted greatly from research in the basic disciplines, especially in biochemistry, psychology, sociology, physics, and chemistry and have exhibited a high level of scholarship in keeping up with home economics research conducted in all parts of the country. Through

the journals that publish the results of investigations, through campus visits of some of the top people in the field, and through travel and advanced study they have kept abreast of national trends and have incorporated into their courses the new knowledge obtained through research."

Branch Stations Enter the Postwar Era

Union

Beginning in 1955, there was considerable cooperation between the Union Branch Station and the main campus on progeny testing of Hampshire and Suffolk rams and on bulls of three lines of Hereford cattle from Corvallis. Range and forestry program work went on as well. For example, the Hall Ranch was used as a field laboratory for range classes from OSC. From 1960 to 1972 almost two million board feet of timber were harvested along with 1,000 Christmas trees.

The Soils, Farm Crops, and Horticulture departments also worked cooperatively with the Union Branch Station in the postwar era in an advisory and planning capacity: chemically analyzing soil and vegetative samples; supplying seed for grass, alfalfa, vegetable, and other plant variety trials; and helping purchase equipment.

Pendleton

The period both before and after World War II saw the successful breeding, crossing, selecting, testing, and screening of many varieties of wheat at the Pendleton Branch Station. This included the "world collection" of 13,000 varieties, each of which had certain qualities superior to its predecessors.

At the time the Pendleton Station was established in 1927, the principal varieties of wheat grown in eastern Oregon were Turkey, Gold Coin, Hybrid 128, and Federation. (Federation was tested and released by the Sherman Station and the USDA.)

In the early 1930's, the new variety Rex was selected, tested, and released jointly by the Pendleton and Sherman stations. It yielded about three bushels an acre more than Federation and was highly resistant to smut. The milling quality of Rex proved to be rather poor, however, so a search for a variety more acceptable to millers continued.

In 1932, the Station was instrumental in the selection of Alicel. This variety never became important commercially, but it eventually led to the well-known Elgin, which, by 1949, had become the leading variety grown in eastern Oregon, yielding about three bushels per acre more than Rex.

Elgin was replaced by Elmar, which was developed and tested by the Washington Experiment Station and released after being tested and screened at the Pendleton Station. At the time of its release, Elmar was resistant to the prevalent strains of smut in eastern Oregon, which had plagued Elgin. Later, however, Elmar became susceptible to other smut strains.

Omar came next and was tested at Pendleton. It resisted all prevalent smut strains. Gaines succeeded Omar and provided a yield increase of six to eight bushels per acre in some areas. It was jointly released from the Washington, Idaho, and Oregon Experiment Stations after testing and screening in these areas.

Moro was developed by the Pendleton Station and released in 1965. It proved to be resistant to smut and also to stripe rust, which was causing devastating losses to wheat growers.

Southern Oregon

Meanwhile, at the Southern Oregon Branch Experiment Station, a number of insects and diseases affecting pears were controlled during the period up to 1965: codling moth (1962), pear leaf aphids (1940), pear psylla (1950), San Jose scale (1935), woolly apple aphid (1930s), spider mites. This research still continues.

Klamath

At the Klamath Branch Station, work during the postwar period included reclaiming saline-sodic soils to productivity, reducing potato losses and increasing yields, and introducing superior cereal varieties of oats and barley. New tests were started on oil-producing crops like flax. Forage crops were improved.

Columbia Basin Combines Sherman and Pendleton

At the Columbia Basin Agricultural Research Center, the old Sherman Branch Station and the Pendleton Station combined, several experiments have been underway since 1931. Such long-term research guides future agricultural developments by identifying the effects of crop rotation, varietal involvement, fertilizers, aerial and surface contaminants, and organic amendments on soil productivity and soil properties. Soil deterioration can be eased or prevented by such work.

Hood River Becomes Mid-Columbia

In 1947, the Oregon Legislature appropriated $20,000 to establish an experimental area in The Dalles for conducting investigations and experimentation into the "problems of producing, fertilizing, harvesting, varietal testing, soil improving, irrigating, handling, storing, utilizing, controlling diseases and pests, and on such other problems of horticultural and field crops as may arise in the area."

The 1953 Legislature changed the name of the Hood River Branch Experiment Station to Mid-Columbia Experiment Station and broadened its scope to include both Hood River and Wasco counties. This action combined the research work of The Dalles Experimental Area and the Hood River Branch Station.

The Mid-Columbia Station continued its record of service to orchardists in the two counties by improving soil fertility and reducing weeds to increase tree growth, studying viruses, and controlling insects and diseases.

Squaw Butte

At the Squaw Butte Branch Station, research begun in the late 1930's continued into the 1950's and 1960's. Work included general range improvement research and the study of the balance between livestock and range forage, the feasibility of water hauling to lead cattle from overgrazed to less used areas, increasing forage by sagebrush control, control of poisonous weeds, improving the winter feeding of range cows, improving wintering of weaner calves, and producing slaughter-grade beef on improved native range for less cost.

The Harney Branch Station, founded in 1911, closed in 1954. The land was returned to Harney County. Its research on natural meadows and livestock production was transferred to the Squaw Butte facility.

Squaw Butte-Harney Branch Station agronomist and Seneca rancher examine a heavy stand of high protein white clover on a July 1, 1954 Experiment Station Field Day. Photo courtesy of OSU Archives, P120:5417.

North Willamette

In 1957, the North Willamette Horticulture Center was established in Aurora, using a $50,000 appropriation by the Oregon Legislature and 52 acres of farm land provided by Clackamas County. Then, as now, the center's goal is to conduct horticultural crop research and to extend new knowledge to the horticulture industry. The center has a diversity of moisture, temperature, and soil conditions offering insight into a variety of growing conditions other than just those in the Corvallis area. Center scientists specialize in research on small fruits, vegetables, and ornamental nursery crops.

Because of the highly competitive nature of horticultural crop production, farmers in other states will try to outcompete Oregon growers; sometimes they succeed. Work at the center has helped farmers by continuing to test and adapt new crops and production systems to keep Oregon competitive.

John Jacob Astor

During the 1950's and early 1960's, research continued at the John Jacob Astor Branch Station in Astoria with good results. The Station's herd of Guernsey breeders was one of 44 in the United States to win a gold star award from the American Guernsey Cattle Club in 1959, 1960, 1961, and 1962. The herd had an outstanding record in two nationwide tests: in 1961, it rated first in butterfat production and sixth in milk production; in 1962, it rated first in butterfat production and second in milk production (in competition with 41,166 Guernsey cows).

The Station carried out research to establish the items that influenced dairy production and determined that good breeding is basic, proper feeding is vital, and effective management is essential. Station researchers also introduced, released, and recommended improved forage crops to dairy farmers. They developed an effective liming and fertilizer program geared to the high rainfall area of the Oregon Coast.

Red Soils

The Red Soils Branch Station in Oregon City had an equally successful research record during the post war period until its close in 1964. Its researchers found practical methods for reclaiming the productivity of Red Hill soils. Formerly unproductive lands were improved so that they could again produce a wide variety of crops. Varieties of good quality grain, legume, grass, and fruit crops were tested, proved, and made available for the growers. Soil-building legumes like red clover and alfalfa proved profitable. Extensive tests determined the economic and effective use of fertilizers. Continuous grain cropping proved inadvisable.

In 1961, Oregon State College officially became Oregon State University.

Malcolm Johnson

by Andy Duncan[3]

Springs of late, Malcolm Johnson has noticed it takes a little longer to chase the soreness from his body when he returns to the fields after the relative inactivity of winter. August seems hotter. Frosty fall mornings chill him deeper.

After nurturing high-desert agriculture for more than 30 years, helping men and women find ways to scratch the most they can from farms he likens to "potholes among the rock ridges," the first and only superintendent ever of the Central Oregon Agricultural Experiment Station at Redmond says he is ready for "a long, long vacation with my wife." Retirement is on his mind.

And when it comes—December 31, although he will work part time until his replacement is hired and oriented—OSU will lose a veteran scientist whose career spans a period of dramatic change in central Oregon agriculture.

"Horses were just going out of use when I got here," he recalls of the February day in 1948 when the Agricultural Experiment Station sent him to Redmond to help farmers adapt to the new strategy in Jefferson County at the northern end of the central Oregon area—irrigated agriculture.

"I had a pair of household scales and a woodshed," he says, smiling as he surveys the memorabilia in his office and remembers how that assignment as supervisor of what then was called the Deschutes Experimental Project grew into the branch station of today, with its buildings, machinery, and research sites at Redmond, Madras, and Powell Butte.

"There was a lot of agriculture when I got here," he continues, "and the people weren't doing all that badly. But after awhile it became apparent they weren't going to be able to get by anymore on a 40-acre farm. It was taking more and more acres to justify the equipment you needed to make a living by farming."

Johnson changed with the times, redirecting his station's research to try and solve farmers' problems as they came along, and he grew fond of his job.

"In those early days the system we used kept me very close to the farmers," he explains. "We did the research right on their land. One of the things I liked about that was getting to know people as individuals, and it kept you in close contact with their problems."

But there were drawbacks, too.

"The big hitch, of course, was that you couldn't do any long-term or specialized research on a farmer's land," he says, pointing out why the Branch Station later acquired property for research.

Johnson, his colleagues say, eventually constructed a program that combined personal problem-solving with other types of research—but with main emphasis always on meeting farmers' needs. By 1957, when he took 2½ years off to go to the Midwest and earn a doctorate in physiology and ecology ("glorified crop production," he says) to add to his bachelor's and master's degrees in agriculture from OSU, his professional path was firmly set.

Returning to Redmond, Johnson slipped back into the quiet, mild-mannered, reliable stance that has earned him respect and gratitude from farmers, fellow scientists, and administrators, including [then] current Experiment Station Director John R. Davis, the last of several bosses who have evaluated his performance.

"We're going to miss Malcolm when he retires," says Davis, "because the people of central Oregon rely on him a great deal for new knowledge. When I'm over there it seems they're always swamping him with questions about how to do this or that with those goofy volcanic soils they farm. He's developed a research program that's 'right on' as far as they're concerned."

Davis's comment is typical of what many at OSU are saying these days. But George Carter, superintendent of the branch agricultural research station at Klamath Falls and Johnson's friend for 20 years, offers an appraisal that may cause more people in central Oregon to nod their heads in agreement:

"The philosophy that still water runs deep applies to Malcolm," says Carter. "He doesn't say much. But when he does you better listen, cause there's a lot more there than meets the eye. Those people must hate to lose him."

For Pets and Animals, A New Leash on Life

by Richard K. Floyd[4]

A dog's best friend—and a cat's, too?

It may well be OSU entomologist Robert L. Goulding. Because of him, many of the nation's 25 million dogs and 30 million cats are free from fleas and ticks.

They wear a plastic collar inbedded with Vapona, an organophosphate insecticide, usually effective for three months. Since 1964, when the "flea collar" was made available to the public, more than 70 million have been used on dogs and cats.

Goulding did the initial work at OSU in developing the use of a plastic formulation as a collar pest control system and much of the work leading to its registration.

In 1963, he and his team of scientists took a fly strip (the kind hung from the ceiling and called spaghetti) and taped it around the dog collar. The spaghetti strip released chemicals which killed fleas. The bigger the strip, the bigger the chemical output.

Working with dogs and then cats, Goulding found the system currently available unsatisfactory because the chemical release rate was too high. After determining the efficiency and limits of safety—vital because release of the chemicals depends on temperature of the animal and the air—the team established a safe concentration of plastic for Vapona, a vinylphosphorus pesticide that breaks down rapidly.

Today the collars are made of polyvinyl chloride resin with Vapona mixed in during formulation of the plastic. The material is extruded or injection-molded as a strip.

The collar releases Vapona at a fairly regular rate. It is picked up in the animal's hair coat in the vicinity of the collar, and since fleas and other body insects move a great deal they are continually exposed. The insecticide also is distributed by the collar rubbing against the animal's body.

The insecticide affects the nervous system of fleas and ticks and kills them. The flea population drops sharply in the first 24 hours. There also is a marked effect against lice, usually not a major pet problem.

Collars for cats are similar, with potency of the insecticide release halved. But it works the same.

The research team has developed a mathematical model to provide a pesticide resin formulation for a particular release rate which could vary among animals or climates. It would be used for both pesticides and attractants, the latter to draw the housefly, cockroach, fruit fly, yellow jacket, termite, and other pests to traps where pesticides could be used, another safeguard for humans since it helps control the amount of chemicals released in the air.

Goulding and other scientists are finding out more about attractants and how to use them, but there is no product like the flea collar yet on the market for humans. However, in the southwest United States, scientists are working on ankle strips which would protect against a tiny red scourge, the mighty chigger.

Goulding is continuing work to broaden the use of pesticides and attractants in plastic. The target—agricultural, industrial, and structural pests. In addition to his team of scientists, he has several other helpers—laboratory dogs and cats test collars with new shapes and capabilities.

There also is a unique maternity section on campus. In a small room, its temperature and moisture carefully controlled, a vital link in Goulding's research is produced.

The room hatches fleas.

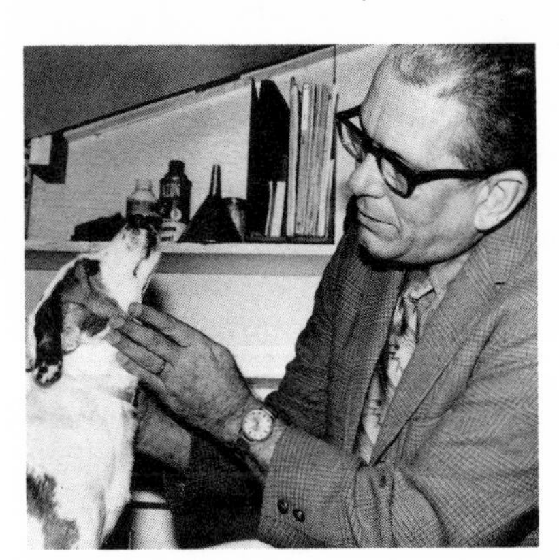

Entomologist Robert L. Goulding consults with a laboratory helper, a dog testing new shapes of flea collars. Goulding developed the plastic flea collar as a means of pet pest control and did much of the work leading to its registration. The collar was made available to the public in 1964.

Oregon Agricultural Research and Advisory Council

The Oregon Agricultural Research and Advisory Council was organized on November 19, 1948 to act as an advisory group representing all segments of Oregon agriculture to the outside world: the State Department of Agriculture, the state budget director, the governor, the legislature, and the Oregon State Board of Higher Education. Board members also spoke for agriculture internally by meeting with officials of Oregon State College, the Extension Service, and the Experiment Station.

Thirty-one organizations belonged to the council, each appointing its own representative. All agricultural producer and processing groups were eligible, subject to a majority vote of members attending any annual meeting.

Members included:

Columbia River Salmon and Tuna Packers Association, Essential Oil Growers League, Northwest Canners and Freezers Association, Nut Growers Society of Oregon and Washington, Oregon Association of Nurserymen, Oregon Association of Soil Conservation Districts, Oregon Broiler Growers Association, Oregon Cattlemen's Association, Oregon Dairy Industries, Oregon Dairymen's Association, Oregon Farm Bureau Federation, Oregon Feed and Seed Dealers Association, Oregon Fryer Commission, Oregon Milk Producers, Inc., Oregon Poultry and Hatchery Association, Oregon Seed Growers League, Oregon State Beekeepers Association, Oregon State Farmers Union, Oregon State Fur Breeders Association, Oregon State Grange, Oregon State Horticultural Society, Oregon Strawberry Council, Oregon Turkey Improvement Association, Oregon Wheat Growers League, Oregon Wildlife Federation, Oregon Wool Growers Association, Pacific Bulb Growers Association, Portland Wool Trade Association, Southwest Oregon Cranberry Club, and Western Oregon Livestock Association.

Marion Weatherford was first president of the council. Over the years, recording secretaries were Ralph Besse, Bob Henderson, and Wilson Foote.

Other members during the existence of the council were: William A. Schoenfeld, F. E. Price, Dean Walker, Carey L. Strome, James A. Doty, H. J. Moulton, Lyle Hammack, Larry Williams, Gordon Walker, Walter Schmedler, Cecil Tully, Leon Fuuke, Claude Williams, and Dudley Sitton.

"Very early in its existence, the council participated in the development of the Experiment Station budget," according to Foote, associate director emeritus. "The council was also instrumental in getting funding for greenhouses, plant pathology research, land for the central station, staff salaries, the College of Veterinary Medicine, and the North Willamette Branch Station. It also helped raise money to build Withycombe Hall, Nash Hall, and the Veterinary Isolation Facility."

A letter from Larry Williams, council president in 1956, to Chancellor John Richards, indicates the way the council operated:[5]

"The Council requested me to convey to you that they regarded the proposed expansions in agricultural research as urgent. Other research needs were discussed but were deferred to those first proposed. The budget requirements to carry out the proposed program were discussed and the Council concluded that they were reasonable. It is the wish of the Council that you and the Board of Higher Education indicate these needs to the Governor and the Legislature when the budget requests for Higher Education are presented. . . ."

"The Council appreciates your interest and understanding of the agricultural activities under your administration. It stands ready to confer with you regarding the research program and budget at your call. The Council also plans to advise with legislators regarding the importance of agricultural research as a wealth-producing activity in our State."

Home Economics Research Accomplishments

The long involvement of the OSU College of Home Economics in research founded by the Oregon Agricultural Experiment Station has resulted in a number of significant accomplishments:

- Helped establish the human requirement for vitamin C
- Developed methods to determine vitamin B-6
- Helped establish human requirement for vitamin B-6 and the indices for determining human status.

- Established that unpaid work done by farm women is not changed when they take on paid employment
- Identified maternal alcohol ingestion with zinc and copper deficiencies and bone and tooth abnormalities in offspring
- Assessed the nutritional status of Oregonians as part of a national study
- Confirmed effectiveness of water flouridation in reducing dental caries
- Developed new recipes for using Oregon's soft wheat
- Showed that three hours is the maximum safe-holding period for perishable foods at room temperatures
- Identified misinformation that causes food loss in households
- Established that families more successfully faced underemployment if their interpretation of the problem did not involve blaming each other
- Matched variety to cooking qualities for the 71 varieties of apples available to Oregon horticulturists in 1915
- Identified in 1926-1927 that the most important factor affecting the use of homemaker's time was the house itself
- Determined desirable work surface heights and storage spaces in homes to meet the needs of families
- Assessed the use of clothing labels by consumers in clothing purchase decisions
- Determined the extent of dermatological health problems tied to clothing, household textiles, and fabric care products
- Confirmed impacts of personal appearance on first impression
- Used time records to measure the work loads of women off and on the farm in 1927 and again in 1977.

Experimental Highlights, 1946-1965

1948 The Oregon Agricultural Research and Advisory Council was organized to act as an advisory group to the Experiment Station. The practice of open-field burning was introduced to control blind seed disease in perennial ryegrass. The Deschutes Experimental Project was established in Redmond to serve farmers in central Oregon; Malcolm J. Johnson was appointed first superintendent.

1949 The Oregon Fish Pellet was developed as a food for hatchery-raised trout and salmon. This pellet now is used worldwide. The Department of Agricultural Economics was established. The wheat variety Elgin was developed at the Pendleton Station and became the leading club wheat variety in the Pacific Northwest.

1951 The first atomic energy grant was obtained by the Department of Agricultural Chemistry for research on the use of radioisotopes as tracers in agricultural chemical research. E. R. Jackman began trials with grazing-type alfalfa varieties.

1952 Studies were initiated to assess the fluoride status of tree fruits in The Dalles before the establishment of the local aluminum processing plant. The North Willamette Experiment Station was established near Aurora; Richard Bullock was appointed first superintendent.

1953 The U.S. Public Health Service transferred to Corvallis a toxicity and fishery unit that later became the Pacific Cooperative Pollution and Fisheries Research Laboratory under the direction of Charles Warren and Peter Doudoroff.

1954 A tri-state (Washington, Idaho, and Oregon) range ecology project was established in eastern Oregon under the leadership of Charles Poulton.

1955 Extensive studies of Vitamin B-12 were initiated.

1958 Pole-frame structures designed by LeRoy Bonnickson became a widely used farm building. Alkali bee beds were established by William Stephen to help pollinate alfalfa seed. The first dwarf apple rootstocks were introduced.

1959 The Department of Animal Science was established, with research programs with animals undergoing a transition from demonstrations to scientific studies. A cooperative research team discovered the cause of "white muscle disease," which annually killed thousands of young calves and lambs, was a deficiency of selenium, a little-known trace element. Fungicide programs to control rust diseases in bluegrass seed fields were developed. Karmex diuron was introduced to control weeds in Willamette Valley wheat fields. Subclover was introduced as a legume for western Oregon hill lands. Ways were developed to propagate the leaf-cutter bee to assist in the pollination of alfalfa seed. The nationally recognized Guernsey herd of the John Jacob Astor Experiment Station won its first gold star award for milk production.

1960 The first colony of cinnabar moths was introduced as a natural control for tansy ragwort, a serious weed problem in western Oregon.

1961 Studies on the causes of brown core in stored pears led to the establishment of controlled-atmosphere storage for pears.

1962 Scientists found that pear decline had become a serious disease in southern Oregon, and research studies began to seek ways to control this disease.

1964 The Willamette tomato, which became the standard western Oregon tomato variety for many years, was introduced.

1965 Control methods for liver flukes in sheep were developed. The OSU 949 bush green bean, one of the first bush beans with Blue Lake quality, was introduced. This variety led the way for the bush bean industry. Pear decline was determined to be caused by mycoplasm organisms transmitted to the trees by psylla insects feeding on the leaves.

Chapter 7 Notes

[1] "The Evolution of the Crop Science Department, 1916-1959," by Don Hill, Corvallis, Oregon, 1985.

[2] Ava Milam and J. Kenneth Mumford in *Adventures of a Home Economist.* Corvallis: Oregon State University Press, 1969. Reprinted with permission.

[3] Andy Duncan, "Malcolm Johnson," *Oregon's Agricultural Progress,* Fall 1980. Oregon Agricultural Experiment Station, Oregon State University, Corvallis, Oregon.

[4] Richard L. Floyd, "For Plants and Animals, a New Leash on Life," *Oregon's Agricultural Progress,* Spring 1971. Oregon Agricultural Experiment Station, Oregon State University, Corvallis, Oregon.

[5] Letter on file at Oregon Agricultural Experiment Station office, Oregon State University, Corvallis, Oregon.

CHAPTER 8

Agricultural Research in the Modern Era (1965-1988)

G. Burton Wood, AES Director[1]

The modern era for the Oregon Agricultural Experiment Station began in 1965 with the selection of G. Burton Wood as director. That same year, Wilbur Cooney became dean of the School of Agriculture. In Wood, the Station had a man who could bridge the gap between the past and present. After all, he had been a student at OAC in the 1930's, and President Kerr's chauffeur. He had known most of the deans of Agriculture and Station directors in the intervening years. He had headed the Department of Agricultural and Resource Economics for 14 years.

Yet, he was very much in tune with the future and the Station's role in bringing its constituents—the farmers and citizens of Oregon—to it.

"As director, I felt I had two functions," says Wood. "One was internal, the other external. It was important to cover both bases." He decided he needed to make sure the branch stations had a close working relationship with the campus departments and vice versa. The results of campus and branch station research then needed to be transmitted to the people who could use them: the farmers and consumers.

Headquarters Staff

Wood set up a headquarters staff to help him.

"Wilson Foote would handle projects, Bob Alexander finance, and Bob Henderson would be the external man," continues Wood. All set about their new tasks. Henderson was soon giving his slide presentation, "The Magic of Research," to audiences all around the state. They were all assistant directors, and Dave Moore eventually joined them.

Wood embarked on a continuing campaign to get away from what he calls "the fount of knowledge" in Corvallis; he frequently visited the branch stations and the many groups and individuals who use and rely on research information.

Robert Mason, another headquarters staff member, assisted him by carrying out periodic surveys of public opinion. "This told us how we were doing," says Wood. "He determined what people thought was most important and least important, what the emerging problems were, what they liked and didn't like about us and our publications. I learned a great deal from this. It gave us a good feel for the public and helped us mold our programs."

From the survey results, Wood picked up the need to think more about the environment and the assessment of technology—in his words, "What happens as a result of the research we do."

"As a result of these studies, we became one of the early states in the West to build up a small cadre of projects relating to environmental quality," he says. "There was a secondary use to these surveys; when I went to the legislature in support of my budget, I didn't go empty-handed. We could keep our dollars going to the most productive areas for the benefit of Oregon agriculture."

National and International Recognition

The years in which Wood was director of the Station brought to the forefront a number of faculty members and branch employees who achieved national and international reputations.

Warren Kronstad is internationally known for his cereal breeding work. He was instrumental in the 1970 University decision to join the Rockefeller Foundation and Centro Internacional de Mejoramiento de Maiz y Trigo (CIMMYT) of El Baton, Mexico to operate a program in Turkey's Wheat Research and Training Center. Today, the cereal imment program with CIMMYT and Agency for International Development includes a strong program on winter-spring crosses of wheat. There are breeding programs in about 100 countries, and all have access to germ plasm maintained at Oregon State through Kronstad's program.

Other Oregon Agricultural Experiment Station staff are internationally known. Walt Mellenthin, superintendent of the Hood River Experiment Station until his December 1982 retirement, is known for his work in pear harvest and postharvest storage work. William Sandine of the Microbiology Department has earned the plaudits of the world's cheesemakers through his work with fermentation bacteria. John Fryer, who heads Microbiology, has developed fish vaccines used around the globe. Harold Evans, a member of the National Academy of Sciences, whose work on nitrogen fixation is known internationally, headed the campus Laboratory for Nitrogen Fixation Research until 1989.

Lowered Funds

Not all recent years have been benevolent to the School of Agriculture. Lowered funding in the 1970's brought about the closing of the Astor Experiment Station, open since 1913 in Astoria. Some of the other stations were combined with the Eastern Oregon Agricultural Research Center and the Columbia Basin Agricultural Research Center. Earlier, the Red Soils Experiment Station in Clackamas County, which operated from 1939 to 1964, was closed because of changing needs.

New Challenges

Typical of the widening interest in research after World War II was the work to help heal scars of three major Tillamook forest fires. H. B. Howell, superintendent of the Astor Experiment Station, was a key worker in seeding, fertilizing, and developing burned-over areas at the Northrup Creek Area. Some of the measures then proposed are just reaching the payoff stage now, and Howell's findings have had implications for much of western Oregon's hill land.

The School of Agriculture recognized early the growing need for marketing research and expanded the Agricultural and Resource Economics Department. There was new interest in learning more about rangelands and continuing the work on selenium, which had been found by University scientists to be the cause of disabling disease in livestock when not present in sufficient quantities.

In the late 1960's and early 1970's, the environment took on new meaning for Oregonians, and the School of Agriculture responded. After much public criticism of the air pollution caused by field burning each year, agricultural engineers built a prototype field burner and several models that destroyed the straw left from grass seed harvest. Machine data were shared with commercial machinery producers. The study of weather and the establishment of burning schedules to alleviate the problems of field burning in the Willamette Valley also were pushed by the school.

As hand labor became more scarce, agricultural engineers tackled the challenge of harvesting strawberries by machine. Food scientists continued to help food processors in their quest for better-tasting foods. Crop scientists began to explore the potential of meadowfoam to produce oil to replace sperm whale oil and other crops that might have a future in parts of the state. The hunt for better varieties of cereals, fruits, berries, seeds, and nuts was kept intense.

Burt Wood has a ready list of accomplishments the Oregon Agricultural Experiment Station achieved during his time as director.

"Many farmers could not be doing what they are doing today without us," he says. "A lot of our work has benefited the consumer. For example, the grass seed industry would not be here without the work of Harry Schoth of Crop Science. The same holds true for Warren Kronstad and the wheat industry.

"Some industries we've not helped, like the strawberry industry. We were never able to build a competitive hops industry despite our work. But poultry—the production of eggs and broilers—was very significant. Our swine research has been significant.

"Another thing that pleases me is the increase of interdisciplinary research. When I came, the Station had some exposure to it. When I left, we had as good a program in interdisciplinary work as anywhere in the country. You bring specialists together in this way—crop scientists, weed people, soil specialists, economists—and solve problems. We couldn't have done our work if not for that."

Richard Floyd's tenure as Station editor (1971-1986) saw an expansion of publications and improvement in its magazine, *Oregon's Agricultural Progress.*

The Davis Years[2]

In 1975, John R. Davis, head of Agricultural Engineering, succeeded Wood as director of the Oregon Agricultural Experiment Station.

"One of the problems of the Oregon Agricultural Experiment Station throughout history is that the agricultural industry is so diverse, with so many important commodities, the Station is spread so thin trying to be all things to all people," says Davis. "One of the things I struggled with was the need to maintain a good base of research and cover all of the commodities, as well as small crops with small potential. Oregon is unique in the diversity of its research."

Davis faced great financial difficulties almost from his first day in office.

"Shortly after I became director in 1975, I faced real budget problems," he says. "In 1977-1978, the economy of the state went downward. At a special legislative session, all agency budgets were reduced, and the Agricultural Experiment Station budget was cut almost 10 percent. It was difficult to make changes in existing programs. We could cut some money simply by not filling vacant positions. But the cuts would create a tough time for everybody, especially for those wanting to build existing programs. They would be able only to manage with what was left."

Davis said that, even with the cuts, he tried to instill in the researchers a sense of the importance of science and technology in agriculture. "I encouraged an attitude of scientific commitment and problem-solving. If we were to bring agriculture into the twenty-first century, we would have to have a strong scientific base. More modern techniques of molecular biology and genetic engineering would have to be introduced, and this was the general direction we went—for example, trying to breed plants with disease resistance built into them. I look back on this kind of thing as one of the high points in my time as director."

The former director considers another high point his work with the various commodity commissions and their support of agricultural research. He also looks with pride on a strengthening of the branch stations. "I always liked to go there," he says. "There are good people there. I also tried to strengthen the ties between the branch stations and departments."

Davis is sold on the Experiment Station idea. "It established an organization to provide local agencies to conduct research in areas where it was not being conducted. In Oregon, it originated as an agency to test fertilizer and soils at a time fertilizer salesmen were going around selling bogus fertilizer. It established tractor standards and fertilizer standards. Now it is the leading organization for doing research for a large number of individuals who cannot do research on their own. With a minimum amount of Federal funding—in Oregon's case 10 percent versus a 50 percent state contribution—the Experiment Station has had a substantial economic impact on the state as a whole."

Changes in the College of Agriculture

As the 1970's melted away, Ernest J. Briskey was named dean of Agriculture, succeeding Cooney, who retired. When Briskey became dean June 1, 1979, decreased Federal and state support for many programs had compelled the reworking of some priorities and the cutting of some activities at farms and branch stations. These reductions were part of cuts being made in government statewide as a consequence of a prolonged recession that had severely affected the state's natural resource based industries.

The school, following recommendations of a review committee, eliminated one level of administration in the Extension Service, merged formerly separate fiscal offices, and combined the communication units of the Extension Service and the Experiment Station.

A new building helped keep the successes coming. In 1981, the Crop Science Building was dedicated.

Also in 1981, the School welcomed a new neighbor, the Northwest Plant Germplasm Repository of the U.S. Department of Agriculture, built on Experiment Station land near Corvallis. It was the first in a series of like facilities to be built across the nation to assure the existence of plant varieties in the future.

In 1983, the School of Agriculture became the College of Agricultural Sciences and the Experiment Station became one of its four divisions, along with the Extension Service, international agriculture, and instruction.

In 1987, Thayne Dutson became director of the Station, succeeding Steven Davis, acting director. Robert Witters had been acting director after John R. Davis left in 1985.

Agricultural Hall of Fame

In 1980, under the leadership of President Robert MacVicar, a cultural and conference center was constructed, entirely with contributions of private donors. Agriculture Dean Briskey championed industry support for the new center, asserting it would become not only a meeting ground for agricultural groups that came to campus, but also a place where non-agricultural groups might more fully appreciate agriculture's contributions. An Agriculture Hall of Fame was established at the center. At the first Agricultural Conference Days in 1980, Marion T. Weatherford, Arlington, and Fred A. Phillips, Keating, were inducted. A year later, inductees were Richard W. Hensel, Klamath Falls; Walter Leth, Salem; Frank Tubbs, Adams; and Norman E. Borlaug, Nobel Peace Prize winner, El Batan, Mexico (honorary member).

In 1982, Claude F. Williams, Prineville; Bob and Jean Nixon, Junction City; Joe Naumes, Medford; and Dudley Sitton, Carlton, joined the Hall of Fame. In subsequent years, a number of other prominent agricultural leaders were named to the Hall of Fame:

1983: Riddell Lage, Hood River; Bill Ross, Jordan Valley; and Edwin Stastny, Malin. 1984: Sam Dement, Myrtle Point; Oscar Haag, Hillsboro; and Howard Smith, Tillamook. 1985: Marjorie Griffin, Vale; William L. Hulse, Dufur; and John C. Scharff, Burns. 1986: E. William Anderson, Lake Oswego; Fred Cholick, Portland; and Richard K. Joyce, Sherwood. 1987: Leonard Kunzman, Salem. 1988: Robert Skinner, Jordan Valley.

Another important part of Agricultural Conference Days were honors paid to other agricultural leaders, leading scientists, researchers and counselors, and to top students in agriculture.

International Projects

Some of the current international work in agriculture is conducted as part of the University's membership in the Consortium for International Development, a nonprofit corporation of 11 western universities that pool their knowledge and expertise to aid developing countries. OSU's contributions, mostly in agriculture, include dry-land agriculture, rangeland resources, rabbit research, cereal breeding, expertise in production of more than 150 commodities, food science and processing, environmental health, seed production and technology, integrated pest management, fisheries, aquaculture and wildlife, and irrigation and water management. In 1987, the Office of International Agriculture became the Office of International Research and Development.

Success Stories

The success stories have also continued. Some of the more recent ones:

•Starter cultures developed by microbiologists have revolutionized part of the cheese-making industry, adding stability to the fermentation process and adding dollars to cheesemakers' income.

•Cereal breeders continue to develop exciting new varieties of wheat and barley to keep Oregon growers competitive in today's markets.

•New vegetable varieties that take advantage of earlier seasons and traits of various areas of the state are changing the way of farming—and eating.

•The secrets of pear storage are being unlocked so consumers can eat pears most of the year.

•Shrimp-processing techniques developed at the Seafoods Laboratory help fishermen obtain more shrimp meat and assure freshness at delivery.

•A fish protein concentrate shows promise of helping underdeveloped countries add protein to diets. The powder is made from hake, a species not fully utilized.

•Fish vaccines are making it possible for hatcheries all over the world to maintain stocks of healthy fish.

•Biological controls are being refined to control disease and insects. Coupled with new management techniques, they save time and money and cut down on the amount of chemical sprays.

•Implants and new nutritional knowledge are producing healthier livestock.

•Virus-free potatoes help growers produce more and better potatoes. The technique of using only virus-free material also has been successful with lily bulbs and will be used on other plants.

Modern Solutions to Complex Problems

The first 100 years of the Oregon Agricultural Experiment Station have been a time of constant service to the citizens of the state. The same pattern has been repeated over and over: whenever people in any segment of Oregon agriculture had a problem, they turned to the scientists and technicians of the Station—either on the main campus or at one of the branches—for a solution. In the process, the men and women of the state who ultimately pay the bills through their taxes benefited.

Some recent Station research success stories prove this point.

Stephens Wheat is a Leader[3]

The soft white wheat, developed by OSU cereals breeder Warren Kronstad and his research team, is the major variety grown in the Pacific Northwest. A conservative estimate shows it has contributed approximately $25 million extra per year from 1981 to 1984 in Oregon alone.

Stephens, as of 1984, represented about 40 percent of the total wheat acreage in the Pacific Northwest, which includes a little more than a million acres in Oregon, 1.3 million acres in Idaho, and 2.6 million acres in Washington. Yields from Stephens of more than 10 percent over other varieties are not uncommon.

"Stephens is unusual because it has the genetic potential to take advantage of favorable growing conditions while still performing well under less-favorable conditions," said Kronstad in 1984. "And the last three years we have had favorable growing conditions."

Stephens, Kronstad admits, also is unusual because it has produced well in a diverse series of locations in the Northwest. As part of the never-ending race to keep up with the demands put on a variety, Kronstad is "tinkering" with the genetics of Stephens to make it more durable.

The No. 1 wheat variety also has more generalized disease tolerance so it may outlast the usual life expectancy of a wheat variety, usually about six years.

"It takes from 10 to 12 years to develop a new wheat variety—and about $300,000 in research investment," says Kronstad, pointing out why researchers must constantly have new varieties in development.

Stephens, named for Dave Stephens, an early cereals breeder at the Sherman Branch Station, was introduced in 1981. The Kronstad group released Hill 81, named for D. D. Hill, former head of the OSU Crop Science Department, in 1983. It is expected to replace some Stephens acreage, particularly in areas where winter-hardiness and tolerance to certain diseases are vital.

What are some of the challenges facing Oregon's wheat industry and researchers?

"We must maintain the high-quality white wheat market that Stephens and other varieties represent while acknowledging the new thrust of developing red, hard winter wheats," says Kronstad. "The red varieties are used to make bread, and housewives in other countries are more and more turning to bread products to feed their families."

In addition to working on winter white and red wheat varieties, Kronstad's team is evaluating genetic material it developed in an effort to produce better spring wheats, both soft and hard.

Warren Kronstad, professor of plant breeding and genetics with the OSU Agricultural Experiment Station for over 25 years, developed a soft-wheat variety called "Stephens," which now accounts for about 75 percent of the one million acres of wheat grown each year in Oregon. Wheat is Oregon's biggest cash crop.

Turkey Research Uses Oregon Feeds[4]

Tom Savage, Oregon State University poultry scientist, is looking for several types of turkeys.

If he finds them, the result will be a boon to Oregon's poultry producers and grain farmers. It'll also be good news for the consumers who eat millions of the birds at Thanksgiving and Christmas and, increasingly, all through the year in specialty products like turkey ham and turkey bologna.

One goal of Savage's genetic research with different breeding lines of turkeys is to develop a strain of birds that producers can raise more economically. He particularly wants a turkey that will put on weight quickly, eating a diet of less-expensive, Oregon-grown feed grains.

"It comes down to trying to look for another marketing avenue for Oregon farmers," says Savage. "My philosophy is, let's evaluate any grain grown in Oregon and see how it does as a poultry feed. Maybe we can reach the point where Oregon-grown turkey means not only grown here but on feed grown in Oregon. It's using turkeys as a vehicle for marketing other Oregon agricultural products."

The state's turkeys, Savage explains, are fed corn and soybeans, expensive items shipped mostly from the Midwest. So far he's tested triticale, a wheat-rye combination that can be grown in eastern Oregon, yellow peas introduced to the Willamette Valley, and fababeans, an experimental crop also under study in the Willamette Valley.

"Fababeans turned out not to be suitable for turkeys. They create leg problems," says Savage. But triticale and yellow peas are potential alternatives to out-of-state turkey feeds.

A variety of triticale called Flora, developed by researcher Matt Kolding at OSU's branch Agricultural Experiment Station at Pendleton for production in the northeastern corner of Oregon, was particularly promising. Turkeys fed the grain did well. And in cooperation with Savage, OSU Foods and Nutrition researcher Zoe Ann Holmes studied the cooking traits of birds fed the grain. They cooked well and tasted good.

"Now," says Savage, "our growers have to start getting the yields that will make the price of triticale and yellow peas go down."

Savage also is studying breeding lines of turkeys to try and help Oregon producers solve a costly problem with infertile eggs, which take up space in the incubator but don't hatch.

In related research, one of his graduate students is using new molecular biology techniques to examine turkey genes and identify the ones that cause some male turkeys to be high-volume sperm producers, a desirable trait.

Also, Savage is studying a muscle disease that affects the quality of turkeys, and he and his graduate students are studying use of the birds as an animal model that could be used in human disease research.

The turkey industry in Oregon has a history of success but has tailed off in recent years. Now it is on the rise again, as it is across the country, Savage points out. A key reason is the birds' adaptable meat is going into new products such as turkey ham.

"The gross farm-gate value of turkeys was up 61 percent in 1985 over the previous year, to $12.1 million," he says. "And this year [1985] poultry consumption (chickens and turkeys) exceeded that of pork nationally."

Electric Fences[5]

The work of the Station even extends to fences.

Home on the range has been a shock for some cows in eastern Oregon the last couple of winters.

The hungry animals were confined in small areas on flood meadows by an easy-to-move electric "super fence" designed in New Zealand, leaving them no choice but to eat meadow plants that had been mowed in the fall and raked into piles.

The cattle came through the winters in good shape, fueling OSU Agricutural Experiment Station researchers' hopes that they have found a way to help ranchers get out of the expensive annual business of cutting and baling flood meadow hay, the conventional winter feed.

A few winters ago, feeding rake-bunched hay to range cattle cost about $18 per head less than feeding baled hay to a control group of cattle, calculated OSU animal scientist Harley Turner. The figure includes the cost of buying the electric fencing and moving it from spot to spot.

The research is part of an effort at the Eastern Oregon Agricultural Research Center, Burns, to help Oregon's $300-million-a-year beef cattle industry. In recent years, ranchers have been squeezed between the recession and other factors that cut beef consumption, and a steady increase in production costs.

Still More Challenges[6]

A few examples of other challenges that lie ahead: Experiment Station scientists on the OSU campus and at branch stations are studying warm-weather variations of "short-duration, close-confinement" cattle grazing that may be highly efficient; others are developing better livestock breeds, using computers to identify for producers better ways of selling livestock, experimenting with ways of harvesting timber without removing land from livestock use, and examining cost-efficient methods of killing undesirable plants that compete with forages for nutrients and water.

Using Chemicals Safely[7]

The work of the Station includes advice on when to spray for pests.

Lots of Oregonians want to protect the state from potentially harmful synthetic chemicals. But some "bug scouts," using techniques pioneered in Oregon by OSU Agricultural Experiment Station researchers, are doing something about it.

In the last few years, insect-scouting services have sprung up in Medford and Hood River to serve the state's apple and pear growers, who produce more than $50 million worth of fruit a year.

Growers can hire a consultant through the services who will monitor pests in their orchards, regularly assessing the potential for fruit damage and outlining possible strategies for dealing with the bugs (including simply letting them alone if there aren't too many).

Growers subscribing to the services are looking for ways to curb the skyrocketing cost of chemical pest control. By keeping tabs on orchard pests, they hope to spray their trees only when absolutely necessary. Also, they hope to avoid spraying trees when the chemicals would wipe out populations of beneficial insects that help keep pests in check.

The approach seems to be working. Most apple and pear growers using the services believe they are saving money and using fewer pesticides, while controlling pests adequately.

Researchers at the Southern Oregon and Mid-Columbia branch stations aren't surprised. For more than 10 years, they've studied insect scouting. The research is part of Integrated Pest Management, IPM, a program that stresses coordinated use of chemical and biological pest control strategies.

"We didn't push. The growers knew we had been studying scouting and came to us," says Pete Westigard, an OSU entomologist who helped set up a scouting service in Medford.

"Just a few years ago they were paying maybe $50 an acre for pesticides. Now some are paying $300 or $400 an acre. You can justify hiring a consultant when you're paying that and have the potential of reducing costs by 50 percent," said Westigard.

IPM techniques such as insect scouting are being applied to other Oregon fruit crops, to vegetable crops in the Willamette Valley, to alfalfa seed production in eastern Oregon's Treasure Valley, and to the control of tansy ragwort, a weed that poisons Oregon livestock.

OSU scientists think the reduction of chemical use and farming costs with IPM has just begun. Wait 10 years and see how it's changed Oregon agriculture, they say.

Tissue-Culturing Potatoes[8]

Maladies of another kind—the viruses infecting potatoes—are the subject of other research.

The parents of many spuds grown in Oregon's $100-million-a-year potato industry start life in a test tube. It's an excellent example of applying so-called "biotechnology" to agriculture.

Using a test-tube process called *tissue-culturing*, Experiment Station crop scientists produce tiny potato "plantlets" that are virtually free of harmful viruses. Seed potato growers sell potatoes they produce with the plantlets to other growers, who cut up those seed potatoes and plant the pieces to produce acres and acres of the top-quality baking and processing potatoes harvested each fall in the Columbia Basin, in central Oregon, and around Ontario and Klamath Falls.

Having a basic seed source that is not contaminated with viruses, and testing to make sure any plants generated from the source material are not contaminated, can reduce potato growers' production costs and improve the quality and yields of their crops.

And what is good for Oregon's potato growers is usually good for Oregon's economy. About three-fourths of the potatoes grown in the state are processed here, dramatically increasing the economic impact of the annual potato crop.

The tissue-culturing system also gives OSU potato researchers a way to quickly "clean up" (rid of viruses) experimental potato varieties they test each year at branch experiment stations at Redmond, Hermiston, Klamath Falls, Ontario, and elsewhere.

That's important, say the researchers, because Oregon, Idaho, and Washington researchers are engaged in a potato-breeding race with scientists in the Midwest and Northeast.

The goal is to produce a better potato variety. Right now, the Pacific Northwest has the upper hand because the Russet Burbank potato, America's favorite for baking and processing, grows beautifully here and not very well in other parts of the country.

But if the Midwest and Northeast, nearer population centers, develop a good baking and processing potato, Oregon and the rest of the Northwest could be in for trouble.

The future holds plenty of challenges besides breeding a better potato.

They include finding more markets for Oregon potatoes and finding less expensive methods for producing, transporting, and planting tissue-culture potato plantlets.

In addition, although test-tube tissue-culturing has given researchers a way of producing plantlets for seed potato production that are free of viruses, scientists still need to find foolproof ways of screening out numerous bacterial and fungal diseases that plague growers and consumers.

New Crops for the 1980's and Beyond[9]

The search for new crops also occupies the time of Station scientists.

Although commercial obstacles remain, OSU researchers have domesticated meadowfoam, a Northwest native wildflower being studied for its oil-producing potential. OSU scientists have released a new variety, Mermaid, to growers and have signed a contract with a cooperative group that will take over production of Mermaid seed.

Experiment Station scientists also are looking at cuphea, another oil seed crop. A native of the southern United States and South and Central America, it could become an oil source to replace coconut oil. The Philippines, the main supplier of coconut oil, is considered an unstable source because of continuing political and economic problems.

Cuphea oil could be used in soaps, detergents, lubricants, and food and medicinal products. Researchers grew it in several parts of Oregon to see how it would fare, with harvested yield the main concern. It is hard to harvest cuphea because the seeds shatter easily and fall to the ground.

Scientists from the Experiment Station, the USDA Agricultural Research Service, and industry (through the Soap and Detergent Association) are partners in a cooperative cuphea research program.

The Station also is looking again at soybeans. Through the years, research has shown that cool night temperatures limit the growth of soybeans. But it's possible that problems can be solved by using varieties from countries with similar climates. Germ plasm (genetic material) from Swedish plant breeders and from a Canadian research program is being tested to see how it reacts to Oregon's climate.

OSU researchers also are continuing work on plants such as rapeseed and pyrethrum, a natural insecticide. The search for alternative crops will never end, because Oregon's farmers operate in a world where markets change constantly.

A Better Life for Farm and City Folks[10]

And what of subjects beyond more traditional farming issues?

Does OSU's Experiment Station study only the obvious—farm production and its economic and health impacts? No. The Station helps support research in a range of social and environmental areas. The goal is to compile information that will help both rural and city folks lead a better life.

Near the beginning of the recession of 1982, an OSU agricultural economist studied the impact on small communities when logging mills shut down.

In one project, OSU anthropologists went into a rural Oregon county and studied the lives, attitudes, and needs of small-scale farmers. That led to a special program for small farmers that was carried out through the OSU Extension Service.

In another project, a sociologist and an agricultural economist examined the attitudes of different groups of people—homeowners, senior citizens, and parents—toward topics such as security and safety. One reason was to find out what services Oregonians want and how they would prefer to pay for them (publicly or privately).

Researchers study the influence of agriculture in various Oregon counties, figuring out how census information can be used to predict the impact that changes in farming and land use will have on employment and the economy.

There are many other examples. Researchers study how rural areas can diversify their economies, developing better statistical methods of assessing community needs and examining migration trends in and out of Oregon.

Much of the social research is done through the Western Rural Development Center, headquartered at OSU. The center coordinates research in 13 Western states.

Resources Natural and Human[11]

Then there's the research the Experiment Station helps fund in OSU's Fisheries and Wildlife Department. From bald eagles to rare deer and valuable fish, the work provides a steady stream of information on creatures that many Oregonians feel enrich life in the state.

The research helps point out how valuable resources like land and timber can be used commercially without endangering priceless fish and wildlife.

The challenges of growing food and fiber for people will never go away. Neither will the need to evaluate the attitudes and needs of the Oregonians who do the growing and consuming, or the need to preserve Oregon's bountiful natural gifts.

Strawberry Varieties[12]

But the basic purpose of the Station is still rooted in more traditional agriculture.

Take strawberries, one of the research programs at the North Willamette Experiment Station at Aurora, just south of Portland.

Station superintendent Lloyd Martin maintains a development pipeline for new strawberry varieties for Oregon growers. He does it in cooperation with Francis Lawrence, a USDA scientist stationed in the Horticulture Department at OSU.

"It's like wheat," says Martin. "You get a good one and it's widely planted. But with time the weaknesses seem to come to the forefront. Growers say a strawberry variety just 'runs out.' Because of diseases or pest problems, they become less vigorous and don't look the way they remember them."

Martin's and Lawrence's job is to make sure a replacement is waiting in the wings.

"It usually takes eight to 10 years from the initial cross for a variety to move through the development system," says Martin. Plants are evaluated at every step along the way and thrown out if found wanting in quality or performance.

In one recent year, Martin, Lawrence, and others at the North Willamette Station planted about 6,000 strawberry seedlings at the station. From them they sought the ones with desirable characteristics.

Potentially, six variety selections may approach release as commercial strawberry varieties, says Martin. But after additional screening, probably no more than two of the six will actually be released, he adds.

"The best-looking one may be one we call 4930," he says. "It has a little bit of a sharp taste when eaten fresh, but it has a good deep, dark color and a high processed quality."

Although Oregon strawberries sold fresh are exceptionally sweet and colorful, most of the state's berries are processed, either quick-frozen or sliced and frozen. Oregon is second in the country in the production of processed strawberries, behind California.

"The thing that makes the competition so strong," says Martin, "is that California grows strawberries for fresh market in huge quantities. But if they get rain, and the quality goes down, they just shunt those off into processing. They have a dual system of marketing, whereas we rely almost entirely on processing. Our quality here in Oregon—flavor and color—is what keeps us in the business.

"But," he adds, "California is getting new varieties that yield higher and have better quality. The competition keeps getting stiffer. We've done surveys of growers, and they believe the most important thing we can do for them is develop new varieties. The Oregon Strawberry Commission provides research support, and most of that support is earmarked for variety development."

Those growers are still asking Martin and Lawrence for a variety as high-yielding and easy to grow as Benton, one of the popular ones grown these days, and as good-tasting as Hood, a variety released in the 1960's, says Martin.

The state of Washington is third behind Oregon and California in producing strawberries for processing.

"I think it's important," Martin says, "that we're now cooperating closely with researchers in Washington to develop varieties for the Northwest. We're getting virus-free breeding material from them and propagating hundreds of plants from it by tissue-culture and sharing the plants so both states can get a quick look at them."

The farm-gate value of Oregon strawberries was about $15.5 million in a recent year. But instate processing adds greatly to the economic impact of that figure.

Predicting Milk Shelf Life[13]

Even milk has not escaped the attention of station scientists.

Floyd Bodyfelt, Extension Service dairy processing specialist, and some of his OSU colleagues have found a way to keep many Oregon consumers from having the miserable experience of encountering bad-tasting milk. When this happens, the main culprits are time and little creatures called psychotrophic, or spoilage, bacteria. Bodyfelt and his coworkers have helped Oregon's $184-million-a-year dairy industry save money and produce some of the highest-quality milk in the country.

The key is a test for predicting milk shelf life which the OSU food scientists developed for the state's dairy processors. They did it while working on an Experiment Station project.

"We call it the Rapid Dye Reduction Test, and everybody benefits from it—consumers, dairy farmers, and processors," said Bodyfelt. He explains that in the 1970's, when the OSU effort got underway, the best test for determining the potential shelf life of milk took 10 to 12 days. "We did a survey and found that only about 20 percent of the processors were satisfied with their products' shelf life," he said.

The test they developed allows processors to check milk for spoilage bacteria in 24 hours or less. If necessary, they then can check raw milk or processing facilities for sources of contamination and make corrections in sanitation and temperature control. With the old test, they'd keep processing contaminated milk for days while waiting for the results.

Spoilage bacteria are not a health hazard, Bodyfelt points out. But they do cause off-flavors.

Only about 20 percent of Oregon processors actually use the test, he says. The most important thing it has done is "focus a lot of attention on the only two sources of spoilage bacteria, the raw milk supply and postpasteurization facilities and equipment," says Bodyfelt.

"We used to say only 10 percent of the fault for early spoilage was due to raw milk and 90 percent came from postpasteurization contamination. Now, because of what we learned developing the test, processors say 25 percent of the keeping quality problem is due to raw milk. There's a lot of emphasis on sanitation and temperature control of raw milk, and more awareness of how to prevent contamination after pasteurization."

The result of the microbiological research is a longer shelf life not only for milk but for other high-value Oregon dairy products like cottage cheese and yogurt. In addition, when processors know that the potential pull date for a batch of milk-based products is going to be earlier than normal, they can ship those products, which are perfectly safe, to a market where they will be consumed quickly.

The OSU food scientists aren't finished.

"We're not completely satisfied with the test," says Bodyfelt. "It is simple, very sensitive, and relatively economical to run. But we're working on a way of making it easier to do."

Even so, "now about 75 to 80 percent of the processors are satisfied with the shelf life of their products," he says, adding that Oregon is now known for producing some of the finest milk in the nation.

A Network of Branch Stations and Other Facilities[14]

The headquarters for the Oregon Agricultural Experiment Station is on the Oregon State University campus in Corvallis. But the Station operates a network of branch research facilities that covers the State.

Scientists permanently assigned to the branch stations do on-the-spot research to solve problems tied to the soils, climates, and other characteristics of various regions. Campus scientists do field experiments at the branch stations, too.

Production research is generally aimed at providing better-quality crops at less cost, which does not necessarily mean more production per acre. Often the quality or form of a product must be changed to meet the demand of domestic or foreign customers, such as the Japanese who want supersweet corn, rather than the standard type usually grown in Oregon.

Researchers also test new technology, such as floating row covers to demonstrate improved control of insect-spread viruses. The covers have proved more effective than the insecticide programs used by most growers. The lightweight fabric cover placed over the growing crop acts as a barrier to aphids and other insects that can infect potatoes, lilies, strawberries, and other crops with a number of viruses.

The branch stations have a strong influence on the economies of their surrounding regions. Here is a brief tour of the branch facilities.

At the North Willamette Agricultural Experiment Station at Aurora, just south of Portland, researchers study ornamental and nursery crops, small fruits, berries, and vegetables.

To the northeast, up the Columbia River at Hood River, scientists at the Mid-Columbia Agricultural Research and Extension Center focus their work on Oregon's high-value pear and cherry crops, as well as fruits like apples and peaches.

Farther east, in the Columbia Basin, are the Hermiston Agricultural Research and Extension Center and the Columbia Basin Agricultural Research Center, with headquarters at Pendleton and a branch at Moro. Researchers at the facilities study irrigated and dryland grains like wheat and barley; irrigated crops like peas, potatoes, alfalfa, and melons; and other new and traditional crops.

Southeast of Pendleton, on the Idaho border, is the Malheur Agricultural Experiment Station at Ontario. It provides research support for farmers growing onions, potatoes, sugar beets, and other crops on irrigated land.

Almost due west from there, at Burns in the heart of Oregon's cattle country, are the headquarters for the Eastern Oregon Agricultural Research Center. Its scientists work on rangeland management and animal production. The center maintains a research station at Union, near La Grande, to serve the ranchers of northeastern Oregon.

Farther west is the Central Oregon Agricultural Experiment Station at Redmond. At the main station, and facilities at Madras and Powell Butte, researchers study mint, grasses, alfalfa, and other crops that grow well in the high desert conditions. At Powell Butte they produce seed for potato variety trials done around the state.

South of there, at Klamath Falls, is the Klamath Agricultural Experiment Station. Its researchers work mostly with potatoes, forages, cereal grains, and pasture management for the livestock in the area.

Farther west, at Medford, is the Southern Oregon Agricultural Experiment Station. Scientists do research with pears, grapes, vegetables, small fruits and other crops.

Those aren't all the off-campus facilities. Experimental farms in the Corvallis area serve as sort of a branch station for the Willamette Valley. The Station also serves the state's fishing industry by operating the OSU Seafoods Laboratory at Astoria and supporting research at the Hatfield Marine Science Center at Newport.

Research isn't the only way the branch experiment station network contributes to the state. The branch stations play an important role in the graduate education of OSU students who go on to make important contributions to Oregon agriculture.

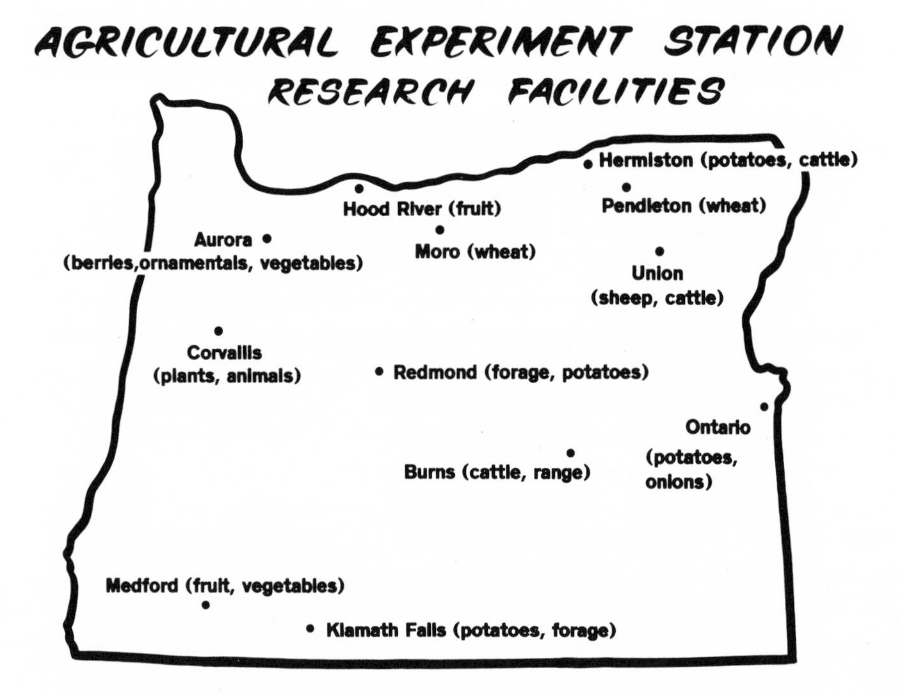

Again, the Hatch Act

"Scientific investigation and experiment respecting the principles and applications of agricultural science."

With these words, the Hatch Act set the Experiment Station idea in motion and authorized money to keep it going. But the Federal Government then stepped back and let the states take it from there.

And the men and women of the Oregon Agricultural Experiment Station—and similar stations in every other state—did indeed take it from there. Over the past 100 years, this dedicated group of scientists and technicians have worked hard—often at great sacrifice—to bring the Experiment Station dream into the magnificent reality it is today.

Congressman Hatch, President Cleveland, and the others responsible for starting it all could only marvel at what their actions in 1888 had produced by 1988.

Experimental Highlights, 1966-1984

1966 Scientists demonstrated that cobalt is necessary for nitrogen-fixating bacteria. Porter Lombard, Peter Westigard, and Clifford Cordy reported on the use of overhead sprinklers for frost protection, pesticide application, heat control, and irrigation of pear trees.

1967 Responding to serious shortages of strawberry pickers, research and development studies were started to develop a mechanical harvester of strawberries.

1970 Responding to excessive air pollution from open-field burning of grass seed fields, research was started to produce a mobile field sanitizer. A vaccine to combat bacterial diseases of hatchery-reared salmon and trout was developed. An agreement was developed that joined the Rockefeller Foundation, OSU, and CIMMYT to operate a wheat production program in Turkey. This became the forerunner for the Station's international programs.

1972 The Western Rural Development Center was established in Corvallis in cooperation with the USDA Cooperative State Research Service and the agricultural experiment stations in the western region. Scientists introduced Cascade hops, a new variety developed from USDA-OSU cooperative breeding programs.

1974 The first dog and cat flea collar, developed by Robert Goulding, was introduced and marketed. The Veterinary Animal Isolation Facility, to prevent the spread of serious animal diseases from research projects, was completed.

1978 The effect of hydrogen fluoride on the fruit set of sweet cherries was demonstrated in The Dalles. Stephens wheat was introduced and became the leading wheat variety in Oregon.

1980 The Harbor Easter lily was introduced for production in southern Oregon. A bacterial start culture was developed to help the cheese industry cut waste and spoilage.

1981 The Northwest Plant Germplasm Repository was established on the Lewis-Brown farm near Corvallis to store the national germ plasm collections of small fruits, pears, filberts, hops, and mint. Scientists pioneered research on the use of natural predators to control serious insect pests in commercial fruit orchards and field crops.

1984 The Agricultural Research Foundation celebrated its 50th anniversary.

Chapter 8 Notes

[1] Ron P. Lovell interview with G. Burton Wood, March 8, 1988.

[2] Ron P. Lovell interview with John R. Davis, April 25, 1988.

[3] Andy Duncan, "What Have We Done For You Lately?" OSU, Agricultural Experiment Station publication, 1981, p. 7.

[4] Andy Duncan, "Where is Oregon Agriculture Going?" OSU, Agricultural Experiment Station publication, 1985, p. 8.

[5] Andy Duncan "What Have We Done For You Lately?" OSU, Agricultural Experiment Station publication, 1981, p. 5.

[6] Andy Duncan "What Have We Done For You Lately?" OSU, Agricultural Experiment Station publication, 1981, p. 5.

[7] Andy Duncan "What Have We Done For You Lately?" OSU, Agricultural Experiment Station publication, 1981, p. 13.

[8] Andy Duncan "What Have We Done For You Lately?" OSU, Agricultural Experiment Station publication, 1981, p. 15.

[9] Andy Duncan "What Have We Done For You Lately?" OSU, Agricultural Experiment Station publication, 1981, p. 17.

[10] Andy Duncan "What Have We Done For You Lately?" OSU, Agricultural Experiment Station publication, 1981, p. 19.

[11] Andy Duncan "What Have We Done For You Lately?" OSU, Agricultural Experiment Station publication, 1981, p. 19.

[12] Andy Duncan "Where is Oregon Agriculture Going?" OSU, Agricultural Experiment Station publication, 1985, p. 12.

[13] Andy Duncan "Where is Oregon Agriculture Going?" OSU, Agricultural Experiment Station publication, 1985, p. 10.

[14] Andy Duncan "Where is Oregon Agriculture Going?" OSU, Agricultural Experiment Station publication, 1985, p. 18.

CHAPTER 9

Patterns of the Past, Partnerships for the Future

By Thayne Dutson,
Oregon Agricultural Experiment Station Director

The Agricultural Research Foundation and the Oregon Agricultural Experiment Station commissioned this history on occasion of the Station's 100 years of service to the State of Oregon. We formally celebrated the Station's birthday July 1, 1988, with ceremonies on the Memorial Union quad at Oregon State University. There, as in this volume, we recognized the many accomplishments of Station scientists, and their numerous contributions to Oregon's economy and citizens.

It is appropriate on such occasions, of course, to afford ourselves the luxury of looking back. While finding satisfaction in the tradition and spirit of service to which our predecessors were so dedicated, those of us affiliated today with the Oregon Agricultural Experiment Station may also benefit from the perspective time provides. From history, we can discern patterns in problems and their solutions. If we are wise enough, we learn from them.

This history is compiled from various sources in campus departments and branch experiment stations. From a careful reading, each of us may distill our own analysis of the patterns of the past. For example, you may derive a new understanding of how Oregon Agricultural Experiment Station scientists have solved problems affecting animals, crops, soils, water, and air, how they have helped open new markets for products from Oregon farms and ranches, how they have long sought new crops as alternatives to those already produced here, and how they have helped protect the natural resources that produce the State's agricultural bounty. It is also clear the State of Oregon values the

Experiment Station as an important player in the economic life of the state. Although the Station has taken its share of budget cuts during economic downswings in the state, Oregonians have consistently chosen to allocate a substantial amount of public monies to advance research in agriculture and natural resources.

History also reminds us that partnerships are essential if the Station is to achieve its objectives and fulfill its mission. A comprehensive list of our historic partners would be long indeed. Many are mentioned in the preceding chapters. Among them are the producers themselves who have contributed in ways too numerous for me to acknowledge here, but who have provided funds, land and equipment for experiments, and valuable counsel about research directions. These producers are also the ones who take the risks of trying a new crop, marketing a new product, or managing new methods.

In a state that grows and markets more than 100 major commodities, commissions and other producer groups have been vital sources of research direction and support, as well as eager recipients and evaluators of the outcomes of our work. Processors, marketers, cooperatives, suppliers of agricultural inputs like fertilizer and chemicals, as well as agribusinesses and financial institutions have been long-time partners.

Another kind of partnership emerged from the far-sighted actions of those who advocated passage of the Hatch Act a century ago. It is the one we enjoy with the federal government and our sister research stations nationwide. Although the federal dollars it provides are important, perhaps of greater value is the interstate network of federal and state research enterprises that has resulted. Much of what is learned through research in Oregon benefits agricultural enterprises and natural resource management in other states. Likewise, this network is a vehicle by which we regularly tap into the talent and knowledge of capable scientists elsewhere by adapting their research to specific situations in Oregon. The value of such cooperation is nowhere more evident than in the cooperative work of scientists in Oregon, Washington, and Idaho—states that share many of the same commodities, climates, soils, markets, and even the same pests and diseases.

There is another, slightly younger but no less important partner that holds a special relationship to the Station: the OSU Extension Service. For more than 75 years, Extension has been the major vehicle for conveying Station research results to people who could apply them directly in practical situations. These Extension agents and specialists are not mere conduits of information, but educators who help people learn to help themselves. Similarly, they in concert with our Branch Station network, link the university to farms, ranches, and communities, providing a sense of critical problems and thereby influencing research directions.

Two more partnerships must be mentioned, because their roles may not be fully recognized, even in this history. One is the Agricultural Research Foundation, formed in 1934 by a group of agricultural leaders who sought to enhance the work of the Station. The Agricultural Research Foundation has provided almost $18 million in research grants since its inception. It also helps Oregonians not directly involved in agriculture better understand and appreciate the scope and direction of our research, by supporting publications and broadcast programs for lay audiences. The Foundation board traditionally has been populated with experienced leaders in agriculture, business, government, and other fields whose wise counsel is reflected in Station policy, research priorities, and research results.

Finally, there is a partnership not often enough recognized, yet essential to the research of any particular time and to the future as well. Of course, I speak of the partnership we enjoy with students, especially graduate students. In our research laboratories, they multiply the effectiveness of our scientists. We rely on them to carry out essential roles in studies designed by our senior faculty. In so doing, we have prepared future generations of scientists who, in turn, have taken their places in laboratories the world over.

It is to all these partners I dedicate this volume. In them rests the future. And it is to the future we should now turn our attention. It may be the prerogative of each generation to believe the problems and the challenges confronting it have never been more complex. Certainly we believe it today, and surely with conviction equalling or surpassing that of scientists and citizens in earlier times. The challenges before us today are unique. My colleague, Extension Director O.E. Smith, points out how the Land Grant university has moved from addressing itself to problems solved by applying a single academic discipline, to problems that required cross-disciplinary solutions, and today to those that demand coordinated effort not only across disciplines, but by numerous institutions and agencies, public and private.

American agriculture is no longer a system of look-alike family farms. Instead, there is a trend toward bimodal agriculture—fewer and larger farms, and many smaller farms. Globally, as individuals and as political-economic units, we recognize the resources of our world are finite and the welfare of the planet rests on our stewardship of those resources. With the advent of high speed, high quality communications and transportation, our world is shrinking. No sector has felt the effects of this interweaving of cultures and economies more than agriculture. On one hand, the presence in world markets of similar commodities from other nations has fueled competition; on the other hand, new market opportunities are present if only we are astute enough to recognize and capitalize on them. We are learning the importance of getting to know our world neighbors better. Oregon may be uniquely well positioned to take advantage of developing Pacific Rim markets for its current and new products.

Domestically, consumer tastes and interests are changing. Concern for personal health, food safety, and environmental protection is prompting people to ask what is in the foods they consume. In reaction, a food system that historically has stressed quality, safety, and low cost is responding, gradually providing foods that are more safe and with even better quality. Our foods are also being produced with safer (and in many cases fewer) inputs such as chemicals and fertilizers. Practices regarding the use of public lands have come under increasing scrutiny of citizens who formerly left such matters to someone else. Agricultural and natural resource managers are learning to deal in a new, wider forum with the issues of public policy governing water, grazing, forestry, toxicology, waste disposal, and similar issues. Resource managers are finding skills in conflict resolution to be as essential as their technical knowledge. Scientists are challenged to explain the benefits of their research and to justify their actions in new forums.

Against this background of change, some realities are remarkably constant. In this history you read of ceaseless challenges my predecessors faced matching resources with need. Almost always they discovered that needs and opportunities for research exceeded available resources. Those same challenges are present today. Because both state and federal governments will be emphasizing fiscal conservatism, our research enterprise can expect relatively level funding. Thus, if we are to shift our priorities, we must do so by deciding not only where to place new emphases, but also where to deemphasize or cut programs. We need to conduct sufficient basic research to allow us to respond to immediate problems while at the same time conduct applied research focused directly on those problems.

Another challenge is one of balancing production-oriented research with that aimed at environmental concerns, natural resource management, marketing, new products, public policy, and consumer interests. We will be tackling problems more and more from a multidisciplinary approach. Teams of scientists will have to work together on today's far reaching challenges, and we need to find ways to reward these scientists for their cooperative efforts. No single experiment station can address all the research interests of its clientele. I predict a continuing emphasis on cooperative regional programs like STEEP (Solutions to Environmental and Economic Problems) with which Oregon, Washington, and Idaho have had so much success containing soil erosion, conserving water, and reducing production costs. We will continue to look to our traditional partners for help in addressing the future. Agricultural producers, processors, and suppliers, as well as state and federal agencies, will continue to provide counsel about needs, as well as gifts, grants, and contracts that will influence directions our research takes.

The paths of Station scientists and Extension Service personnel continue to draw closer. Just as many Corvallis faculty have long held joint Station-Extension appointments, so too will we see more joint appointments of Extension and Experiment Station faculty at off-campus locations, a practice already begun. There will be more co-location of Extension faculty at branch stations and research centers. A major move is already underway at the North Willamette station that will provide better support to the important horticulture industry. Joint administration of both Extension and Experiment Station programs has been accomplished in Southern Oregon at Medford and joint appointment of Experiment Station and Extension faculty has been approved for Central Oregon Experiment Station.

The face of science is changing. Biology was transformed forever when the genetic codes of DNA were unlocked. Principles and techniques established the past decade in molecular biology have laid foundations for the tools we will need to address problems of a scale unknown before. At the same time, the perceived power of biotechnology has drawn attention from outside the scientific community and has hastened debate at the boundaries of science and public policy. Never-before-asked questions arise regarding ownership of new genetic materials. Technology transfer encourages development of spin-off firms and industries. And although we have not, as a society, sorted through the applications of this powerful new tool, I am confident it will have wide application in modern agriculture, allowing us to achieve such emerging objectives as improving food quality and safety while reducing the use of chemical inputs. It may also allow us to produce foods of types before unknown and in places we may have not thought possible.

In remarking on new tools, I cannot overlook the modern computer. More computing power resides on a desktop today than was available in the largest mainframes only a decade or so ago. But what are we to do with this new tool? Surely we can acquire, process, manipulate, and exchange more data and information more quickly than ever before. But to what end? From the past we have observed that the Land Grant university—through its Experiment Station and Extension Service—has made some of its greatest contributions to society by helping interpret information and by bridging the gap between basic science and practical problems. Today, Experiment Station scientists are using powerful computational tools. With their accumulated knowledge of the environment, plants, and animals, they can create models of natural systems that will help us predict and interpret, weigh alternatives, and make informed decisions. We must continue to provide accurate data for input into computer models as well as continue to refine and develop our models.

Even with these marvelous tools of science and technology, our people are still the heart of the Oregon Agricultural Experiment Station research enterprise. Scientists, technicians, support staff, and students: all are critically important. Among them, too, change is taking place. Many of our best, most senior staff who began their careers after World War II and the Korean War are now retiring. They are being replaced by men and women who have grown up with the new science and the new tools and who are vigorous, imaginative, talented, and dedicated, but who often do not have a background that directly links them to farms and ranches. A parallel "changing of the guard" is occurring in ownership and management of farms, ranches, agribusinesses, and cooperative. It is the challenge for those of us in administrative roles to assure that new bridges will be built between Station scientists and those who rely on their research efforts. Likewise, administrators must spread the word of a wealth of opportunities in agriculture and natural resources. Recent U.S. Department of Agriculture reports indicate we may expect a shortage of people to fill the vacancies in agricultural sciences occurring in the next decade. In short, we need more bright students in our laboratories and classrooms. We must be able to assure these men and women that productive, rewarding careers await them.

As the director of the Oregon Agricultural Experiment Station now entering into its second century, I stand in awe of what has been accomplished in its first. But I am persuaded that this state-federal-industry partnership is equally adaptable to the needs and challenges the late 20th and the 21st century will bring us. Its accomplishments will be magnified in the next 100 years!

Appendix A

Directors of the Oregon Agricultural Experiment Station

Edgar Grimm—1888-1890
Benjamin L. Arnold—1890-1892
John M. Bloss—1892-1896
H. B. Miller—1896-1897
Thomas M. Gatch—1897-1907
William Jasper Kerr—1907
James Withycombe—1908-1914
A. B. Cordley—1914-1920
James T. Jardine—1920-1931
William A. Schoenfeld—1931-1950
F. E. Price—1950-1965
G. Burton Wood—1966-1975
John R. Davis—1975-1985
Robert E. Witters—1985-1986, acting director
Steven L. Davis—January-September 1987, acting director
Thayne Dutson—1987-

Appendix B

Current Officials of the Oregon Agricultural Experiment Station

Thayne Dutson, Director
Kelvin Koong, Associate Director
Van Volk, Associate Director
Margy Woodburn, Associate Director

Bibliography

Manuscript and Unpublished Sources

Agricultural and Resource Economics Department history. Oregon State University, Corvallis, Oregon, 1982.

Agricultural Chemistry Department history. Oregon State University, Corvallis, Oregon, 1965.

Agricultural Engineering Department history. Oregon State University, Corvallis, Oregon, 1967.

Agricultural Experiment Station Council Meeting Minutes, 1890-1900. Director's Office, Oregon State University Agricultural Experiment Station, Corvallis, Oregon.

Animal Science Department history. See Rangeland Resource Department history.

College of Agriculture. See School of Agriculture.

Columbia Basin Research Center, Oregon State University, USDA-Agricultural Research Service, Columbia Plateau Conservation Research Center, Pendleton, Oregon, background information.

Crop Science Department history. See Farm Crops history.

Entomology Department history. Oregon State University, Corvallis, Oregon, 1988.

Eastern Oregon Experiment Station (Union) history. Oregon State University Agricultural Experiment Station, Corvallis, Oregon, 1971.

Farm Crops Department history. Oregon State University, Corvallis, Oregon, 1967.

Fisheries and Wildlife history. Oregon State University, Corvallis, Oregon 1978.

Floyd, Richard. "History of the Oregon State University College of Agriculture." Agricultural Communications, Oregon State University, Corvallis, Oregon, 1983.

Food Science and Technology Department history. Oregon State University, Corvallis, Oregon, 1967.

Food Science and Technology Department, "Background Information," in: report to Cooperative State Research Service, USDA, 1985.

Home Economics history. College of Home Economics, Oregon State University, Corvallis, Oregon.

"Horticulture at Oregon State, A Historical Sketch," by A. N. Roberts, 1982.

Horticulture Department history. Oregon State University, Corvallis, Oregon, 1982.

Jackman, E. R. "Wheat in Oregon." Wheat Industry Conference Committee Reports, Feb. 19-20, 1957, Portland, Oregon, pp.1-7.

North Willamette Horticulture Center, Oregon Agricultural Experiment Station, Aurora, Oregon, "About the Station," 1987.

Poultry Science Department history. Oregon State University, Corvallis, Oregon, 1967.

Poultry Science Department Program. Oregon State University, Corvallis, Oregon by G.H. Arscott, 1987.

Poultry Science Department history. Oregon State University, Corvallis, Oregon, 1982.

School of Agriculture and Agricultural Experiment Station of Oregon State University History (A brief). R.S. Besse, 1966.

Soil Science Department history. Oregon State University, Corvallis, Oregon, 1967.

Personal Interviews

Davis, Jack. Interview with Ron P. Lovell, April 25, 1988.

Wood, G. Burton. Interview with Ron P. Lovell, March 8, 1988.

Published Sources

Agricultural Research Foundation 1934-1984. Oregon State University, 1984.

Duncan, Andy. "Malcolm Johnson," *Oregon's Agricultural Progress,* Fall 1980.

"Fifty Years of Research at the Sherman Experiment Station." Oregon Agricultural Experiment Station Miscellaneous Paper 104, June 1961.

Floyd, Richard. "For Plants and Animals, a New Leash on Life," *Oregon's Agricultural Progress,* Spring 1971.

Groshong, James W. *The Making of a University,* 1868-1968. Corvallis: Oregon State University, 1968.

"John Jacob Astor Experiment Station, Its Development, Program, and Accomplishments, 1913 to 1963." Agricultural Experiment Station Special Report 157, Oregon State University, Corvallis, July 1963.

Kerr, Norwood Allen. *The Legacy, A Centennial History of the State Agricultural Experiment Stations, 1887-1987.* Columbia, Missouri: Missouri Agricultural Experiment Station, 1987.

"Klamath Experiment Station, Its Development, Program, and Accomplishments, 1937-1963." Oregon Agricultural Experiment Station Special Report 174, April 1964.

"Mid-Columbia Experiment Station, Its Development, Program, and Accomplishments, 1913 to 1965." Oregon Agricultural Experiment Station Special Report, March 1966.

Milam, Ava and J. Kenneth Mumford. *Adventures of a Home Economist*. Corvallis: Oregon State University Press, 1969.

Oregon State University Bulletin General Catalog. 1988-1990, No. 215, Spring 1959.

"Pendleton Experiment Station, Its Development, Program, and Accomplishments, 1928 to 1966. Oregon Agricultural Experiment Station Special Report 233, March 1967.

"Red Soils Experiment Station, Its Development, Program, and Accomplishments, 1939-1964." Oregon Agricultural Experiment Station Special Report 202, December 1965.

"Southern Oregon Branch Experiment Station, Its Development, Program, and Accomplishments, 1911 to 1962." Oregon Agricultural Experiment Station Special Report 156, July 1963.

"Squaw Butte Experiment Station: Its Development, Program, and Accomplishments, 1935-1969." Oregon Agricultural Experiment Station Special Report 599, Sept. 1980.

"The First Fifty Years of the Oregon Agricultural Experiment Station, 1887-1937." Station Circular 125, Oregon State College, Corvallis, August, 1937.

Turnbull, George. *Governors of Oregon*. Portland, Oregon: Binford and Mort, 1959.

"Umatilla Experiment Station, Its Development, Program, and Accomplishments, 1909 to 1969." Oregon Agricultural Experiment Station Special Report 312, Sept. 1970.

Index

A

B

C

D

G

H

M

N

O

P

R

S

OREGON STATE UNIVERSITY